Red Sea Geothermal Provinces

Red Sea Geothermal Provinces

D. Chandrasekharam, Aref Lashin,
Nassir Al Arifi & Abdulaziz M. Al-Bassam

CRC Press
Taylor & Francis Group
Boca Raton London New York Leiden

CRC Press is an imprint of the
Taylor & Francis Group, an **informa** business

A BALKEMA BOOK

CRC Press/Balkema is an imprint of the Taylor & Francis Group, an informa business

Typeset by MPS Limited, Chennai, India
Printed and bound in Great Britain by CPI Group (UK) Ltd, Croydon, CR0 4YY

Library of Congress Cataloging-in-Publication Data

CIP data applied for

Published by: CRC Press/Balkema
P.O. Box 11320, 2301 EH Leiden, The Netherlands
e-mail: Pub.NL@taylorandfrancis.com
www.crcpress.com – www.taylorandfrancis.com

ISBN: 978-1-138-02696-4 (Hbk)
ISBN: 978-1-315-74739-2 (eBook)

Printed and bound in the United States of America by
Edwards Brothers Malloy on sustainably sourced paper

Table of contents

Foreword

Under the present context of global warming due to excess emissions of greenhouse gases (GHG), countries should give higher priority to renewable energy. The renewable energy technologies will help poor countries to achieve the benefits of development without having to face the same environmental costs the developed world has experienced. Energy and water are becoming the most sought commodities in the growing world today. Every country is trying hard to secure energy, water and food. This is true with respect to countries around the Red Sea. These countries have abundant geothermal resources that are lying untapped. Both oil rich countries, like Saudi Arabia, and oil importing countries like Djibouti, Ethiopia and Eritrea have large geothermal resources. Today the economic growth of these oil-importing countries depends on other countries. Such countries are not energy independent nor energy secured. Furthermore, a common problem with all countries around the Red Sea is the availability of fresh water. Shortage of freshwater leads to food insecurity. Among the Red Sea countries, Saudi Arabia emits 4 billion tons of CO_2/y and is the highest among the countries around the Red Sea. In addition, 13 Mt of CO_2 is emitted from desalination plants. The effects of high CO_2 emissions are already being experienced by the country in the form of increased ambient temperature over the past decade.

The authors in this book brought out excellent mitigation strategies to circumvent such CO_2 related issues by highlighting the valuable hydrothermal and Enhanced Geothermal resources in the countries around the Red Sea. The technology to generate electricity, produce fresh water, support greenhouse cultivation, preserve food products through dehydration through geothermal energy resources is well developed. The advantage of geothermal energy source is that this source can supply baseload electricity unlike other renewables and is on line 90% of the time with least CO_2 emissions.

I congratulate the authors for having produced this book at a time when it is most needed. It is a feeling of gratitude and pride that I find such a book co-authored by two King Saud University faculty members. This, if anything shows how our researchers are concerned with not only problems of their neighboring communities, but also problems of the region in order to secure non-traditional energy resources to secure life. This book, I am sure, will be very useful to graduate students, researchers and institutes involved in renewable energy research. Finally, I recommend that the authors initiate a scientific research program on a regional scale to assess the feasibility of energy emerging from hot spots of the Red Sea and how countries can operate to implement and best be benefited from such energy for the welfare of peoples.

Dr. Abdul-Aziz A.R. Alothman
Vice President for Educational and Academic Affairs
King Saud University, Riyadh, Saudi Arabia

Authors' Preface

"Today, over two billion people in developing countries live without any electricity. They lead lives of misery, walking miles every day for water and firewood, just to survive. What if there was an existing, viable technology, that when developed to its highest potential could increase everyone's standard of living, cut fossil fuel demand and the resultant pollution" said Peter Meisen, President, Global Energy Network Institute in 1997. After seven years the situation remained the same. In the annual meetings held in 2004, the World Bank President said: "We must give higher priority to renewable energy. New and clean technologies can allow the poor to achieve the benefits of development without having to face the same environmental costs that the developed world has experienced". Even after ten years the situation has not changed in the rural sector of the developing countries. 90% of the world population living in rural areas in developing countries have no access to basic needs like nutrition, warmth, and light, in spite of the fact that technologies for developing renewable energies have jumped in leaps and bounds and this is especially so with respect to geothermal energy.

This is especially true with respect to the countries around the Red Sea. It is very interesting to observe that geothermal resources are available in the countries rich in oil and gas resources and in those countries that have no oil or gas resources. Those countries that have rich geothermal resources like Eritrea, Ethiopia, Djibouti, and Republic of Yemen, depend on imported fossil fuels to support their energy demand, instead of developing naturally available geothermal resources. Thus, their economic growth lies outside their country's boundaries. In other words, someone else dictates your economic growth. These countries have no energy independence or energy security. Besides energy, these countries face acute shortages of fresh water because they receive scanty rainfall. Thus, these countries have no food security either. On the other hand, Saudi Arabia has surplus oil and gas resources and is not aware of the vast geothermal resources existing within the Saudi Arabian shield. Among the countries around the Red Sea, Saudi Arabia emits 4 billion tonnes of CO_2 per year. Because Saudi Arabia has surplus fossil fuel reserves, the country is generating fresh water through desalination to support the demand from domestic and agricultural sectors at the cost of emitting about 13 Mt of CO_2. The country is already experiencing a change of weather patterns and an increase in ambient temperature. Due to the fresh water stress status, the country has adopted a policy to stop wheat production from 2016 onwards, which puts the country's food security in other countries; this is a matter of serious concern. Thus, both rich and poor countries have energy assets that have not been utilised as well as one common problem, which is how to supply fresh water for millions of their citizens. All the countries around the Red Sea have excellent hydrothermal and enhanced geothermal systems (EGS) sources that can be developed. The technology to generate electricity from both these sources is mature and available off the shelf. Unlike other renewables, geothermal energy can supply base load power and >90% online. In addition to power, this energy can support greenhouse cultivation, and dehydration of agricultural products. The most significant contribution that geothermal energy can make to these countries is to generate fresh water to meet the growing demand from domestic and agricultural sectors. Countries can be energy independent and can have food security and support the development of rural regions. In the next century these countries can become energy and water surplus countries and support other neighbouring countries to elevate the socio-economic status of the rural population. The book starts with an overview of the carbon intensity index under 6D, 4D, and 2D scenarios the world is currently debating and highlights the levelised cost of electricity generated by a variety of renewable energy sources like solar power, wind, biomass, and geothermal power. The facts are visible to anyone who wants to judge the credibility of geothermal energy. All good things come with a premium, so does geothermal energy. The most important message the authors make to the world community in the beginning is that energy and water dictate the future of any country, whether rich or poor.

Electricity markets, future electricity demands, GDP growth, and population growth in the countries around the Red Sea in the context of the global electricity demand, especially the OECD countries is discussed in Chapter Two. The statement made by Pierre Gadonneix in 2006 is very true. Fossil fuels and renewables have grown together and will do so in the future. At least until 2040 these energy sources will not cross each other's paths. This parallel growth is a sign and a ray of hope, to see the light at the end of this CO_2 tunnel.

Over 69% of CO_2 is emitted by the energy source we use – fossil fuels. It is very obvious that we will cross the 2D threshold very soon if we do not realise how much CO_2 we are emitting. Poor countries are emitting CO_2 since they feel that they have no other options and rich countries – out of greed. However, both have a single option to control CO_2 emissions and save the global climate for our future generations. All countries should honour their commitments in controlling CO_2 emissions. Chapter Three unfolds the CO_2 facts of all the countries around the Red Sea and their contributions to global climate change.

"Geothermal is 100 per cent indigenous, environmentally friendly and a technology that has been underutilised for too long". This statement made by Achim Steiner, UN Undersecretary General and United Nations Environment Programme (UNEP) Executive Director, at the UN climate convention conference in Poland in 2008 is absolutely true, especially with respect to the countries described in this book. Chapter Four unfolds the evolution of the geothermal systems in these countries in relation to the geology and tectonics and gives in-depth knowledge about the quantum of energy resources that are waiting to be explored and exploited. Estimates of electricity that can be generated from the existing hydrothermal systems are highlighted. Chapters Three and Four unfold strategies that the countries can adopt to modify their energy policies to give a better life to their future generations.

The International Energy Agency in 2014 brought out five key actions to achieve a low carbon energy sector. Strong policies are required to decarbonise the electricity sector. CO_2 reduction strategies the countries can adopt, in the light of the debate that is being held between the Conference of the Parties (COP 21) currently, without sacrificing socio-economic growth, are focused on in Chapter Five.

This book is aimed not only at the policy makers but also looks at the young people who are keen on taking geothermal energy as their career option. There is a shortage of trained manpower for developing geothermal energy resources in the countries around the Red Sea. The 'country update' papers in the World Geothermal Congress 2015 reveal that manpower allocation for developing geothermal energy resources is far below the required number. There is lack of, and urgent need for, an easily understandable and accessible information source in a comprehensive form. Chapter Six elaborates the exploration techniques that are generally followed by *geothermally* advanced countries for the benefit of the young graduates who wish to take up a career in the geothermal industries. Textbooks, which focus on regional resources and problems associated with them are very few and for the Red Sea there are none. Although such information does exist in several published scientific papers, it very often becomes difficult for the graduate students and researchers to get such information within a short period of time. This is especially true in the case of students in the above countries, who face difficulties in accessing such information.

Methodologies that are being adopted to generate power from geothermal energy across the world are described in Chapter Seven. Although these techniques are known, inclusion of such details will make the book complete.

"The development of greenhouse agriculture and geothermal based aquaculture in my country also demonstrate how sustainable energy can increase food production considerably, giving farmers and fishermen new ways to earn a living" states Ólafur Ragnar Grímsson, the President of Iceland. It is true; low-enthalpy geothermal energy resources are put to use in a variety of ways to enhance a country's GDP. A substantial amount of cost can be saved on energy bills if geothermal energy sources are used for space cooling and heating, green house cultivation, dehydration and other similar applications. If an agriculturalist in Guatemala could capture the (geothermally) dehydrated vegetables market in Europe, why can't a young entrepreneur in the countries around the Red Sea do it? This would drastically reduce dependency on imported food. Chapter Eight highlights the direct use of geothermal energy sources.

D.W. Brown, D.V. Duchane, G. Heiken and V. T. Thomas, the authors of the book *Mining the Earth's heat: Hot dry rock geothermal energy,* written in 2012, dedicated their book "to those who labored over many years to take the hot dry rock concept from simply a novel idea to a proven reality". Indeed, this technology, now christened as EGS, is being taken up by several countries. Harnessing the Earth's heat for a future better life is the best option we can adopt to mitigate climate change issues and provide sustainable and continuous energy and water to the millions living on earth. This energy not only can generate power but can also support a variety of sectors like agriculture, greenhouse cultivation, aquaculture, space cooling and heating, refrigeration and dehydration. This energy initially may supplement growing energy and water demand that fossil fuels cannot support due to large CO_2 emissions, however in future it may take over the entire fossil fuels trajectory that is shown in Figure 2.3. Coal and oil energy sources are becoming messy although they are easily accessible due to mature technology. The fact is, only a small part of the fossil fuel resources is extracted from the reservoirs and the oil companies are trying hard, by investing huge amounts of money, to enhance oil recovery from the reservoirs. These companies are obsessed with increasing the small percent of oil recovery to get returns on their investment. However, what the companies fail to understand is that there is a huge treasure that is lying within the drilled wells that were abandoned or closed. These abandoned wells, with high bottom hole temperature, are indeed a treasure for the company and the country as well.

Back in 347 AD, the Chinese used drilling bits attached to bamboo poles for extracting oil from about 240 m below the ground. Since then the oil industry has grown in leaps and bounds and reached the present status. Thus technology grows with time and experience. EGS technology is in its initial stages now. However, compared to the early days, the present day technological development is far superior and faster. It will not be too long before the population can have their own EGS well in their backyard. Chapter Nine reveals how much of energy is locked-up in the granites rich in uranium and thorium in the crust around the Red Sea. Between the time when the hot dry rock project was conceived and now, the thinking behind the circulation fluid for heat mining has changed. Currently technology is being developed to use carbon dioxide as a working fluid in EGS systems instead of water. When this technology matures, both thermal power plants and EGS based power plants can co-exists. One becomes a donor of and the other becomes a consumer of CO_2.

Countries invest large amounts of money to extract oil and gas and nearly double that investment is made in cleaning the environment. More than two billion people in the rural regions of the developing countries have no access to electricity, they walk several kilometres to fetch water and fire wood. This costs energy and time, which should also be factored in to the investment. Today geothermal energy comes with a premium since the energy is clean, supply is 24×7 with less down time and the power plants can operate with efficiency ease for 30 to 40 years. The unit cost of electricity generated from geothermal energy source is far less, compared to other renewables. Chapter Ten discusses the economics of geothermal energy both hydrothermal and EGS based power generation.

This is the first of its kind written for the countries around the Red Sea, keeping in mind the graduate students, research scholars, policy makers, financiers and energy planners, who are the backbone of development for these countries. This book is a huge source of information useful for decision and policy makers and administrative leaders. Development banks and financial donors, such as the World Bank, Organisation of Petroleum Exporting Countries (OPEC), Gulf Cooperation Council (GCC), and the African Development Bank, can draw on large information, both technical and economical, to tune their decisions related to financial support to develop geothermal energy source in these countries.

Finally, the authors are confident that this book will be useful for many people helping society to effectively use the huge available geothermal resources, providing energy and water security and energy independence to the rich and the poor countries around the Red Sea.

Dornadula Chandrasekharam
Aref Lashin
Nassir Al-Arifi
Abdulaziz M. Al-Bassam

About the authors

Dornadula Chandrasekharam (born 1948, Chandra, India), Chair Professor in the Department of Earth Sciences, Indian Institute of Technology Bombay (IITB), obtained his MSc in Applied Geology (1972) and PhD (1980) from IITB. He has been working in the fields of geothermal energy resources, volcanology, and groundwater pollution for the past 35 years. Before joining IITB, he worked as a Senior Scientist at the Centre for Water Resources Development and Management, and Centre for Earth Science Studies, Kerala, India for seven years. He held several important positions during his academic and research career. He was a Third World Academy of Sciences (TWAS, Trieste, Italy) Visiting Professor to Sanaa University, Yemen Republic between 1996–2001; Senior Associate of Abdus Salam International Centre for Theoretical Physics, Trieste, Italy from 2002–2007; Adjunct Professor, China University of Geosciences, Wuhan from 2011–2012, Visiting Professor to King Saud University of Saudi Arabia (2014–) and Adjunct Professor, University of Southern Queensland, Australia (2016–2019). He received the International Centre for Theoretical Physics (ICTP, Trieste, Italy). Fellowship to conduct research at the Italian National Science Academy (CNR) in 1997. Prof. Chandra extensively conducted research in low-enthalpy geothermal resources in India and is currently the Chairman of M/s GeoSyndicate Power Private Ltd., the only geothermal company in India. He is an elected board member of the International Geothermal Association, and has widely represented the country in several international geothermal conferences. He conducted short courses on low-enthalpy geothermal resources in Argentina, Costa Rica, Poland, and China. He has supervised 25 PhD students and published 160 papers in peer reviewed journals of international repute, and published seven books in the field of groundwater pollution and geothermal energy resources. His three books on geothermal energy resources are: 1) *Geothermal Energy Resources for Developing Countries* Balkema Publishers (2002), 2) *Low Enthalpy Geothermal Resources for Power Generation*, Taylor and Francis (2008), and 3) *Geothermal Systems and Energy Resources Aegean and Eastern Mediterranean Region*, Taylor and Francis (2014), are widely read. Prof. Chandra served on the Board of Directors of: 1) Oil and Natural Gas Corporation, 2) Western Coal Fields Ltd., 3) India Rare Earths Ltd., and 4) Mangalore Refineries and Petrochemicals. He has been appointed as the Chairperson of the Geothermal Energy Resources and Management Committee constituted by the Department of Sciences and Technology, Government of India.

Aref Lashin, currently working as an Associate Professor in the Department of Petroleum and Gas Engineering, King Saud University, Riyadh, obtained his PhD from the Freiberg University of Mining and Technology. His permanent affiliation is with the Department of Geology and Geophysics, Faculty of Science, Benha University, Egypt. He was a graduate student at the United Nations University, Geothermal Programme, Iceland, in 2005. Dr Lashin's expertise is in the fields of geothermal energy and hydrocarbon exploration and he is currently conducting several geothermal projects supported by the National Science, Technology and Innovation Plan (NSTIP) strategic technologies programmes in the Kingdom of Saudi Arabia. He has published a large number of scientific papers in ISI and peer reviewed journals and supervised many masters and PhD students. Dr Lashin can be reached through his regular email: aref70@hotmail.com.

Nassir S. Al-Arifi, PhD from the University of Manchester, UK, is currently the director of the visiting professor programme, vice rectorate for scientific research at King Saud University and a full professor specialised in earthquake seismology at the Department of Geology and Geophysics, College of Science, King Saud University, Riyadh. Professor Al-Arifi published more than 45 articles in peer reviewed journals and supervised more than 17 MSc theses. His current research interest is in the field of geothermal exploration using multi method applied geophysics; he has published more than 15 articles in this field. He translated two books in the fields of hydrogeology and geophysics.

Abdulaziz M. Al-Bassam is a Professor of hydrogeology and hydrochemistry, Department of Geology, College of Sciences, King Saud University. He obtained his MSc degree from Ohio University, USA and PhD from Birmingham University, UK. Earlier, he was the Chair of the Geology Department, Vice Dean for the College Sciences, King Saud University, and member of the board for the Saudi Geological Survey (SGS). He held several important positions in the Kingdom: Secretary of Saudi Society for Earth Science; Supervisor of Saudi Geological Survey Research Chair (SGSRC) for Natural Hazards; Member of Advisory Committee for Prince Sultan Bin Abdulaziz International Prize for Water; Member of Executive Committee for Dams Operation and Maintenance **Program.** His research interests are hydrogeology, hydrochemistry, geothermal and environmental aspects. He has published more than 50 papers, authored two books and translated one book.

Acknowledgements

The Red Sea rift is one of the tectonically active regions on Earth. When this is combined with volcanic activity, the regions become geothermal potential regions. Several discussions held during DC's tenure as Visiting Professor in King Saud University on the above subject culminated in the idea of bringing out a book on this topic. Geothermal energy, being the most important renewable energy, is abundant in all the landmasses around the Red Sea, but there is no book available for students of the region to understand and peruse research in this field. This idea has now become a reality for the benefit of a wide section of the student and research community. To bring out a comprehensive book of this type needs support from various sections of the academic, administrative, and research communities. Our sincere thanks go to the Office of the Deanship of King Saud University (KSU) and the administrative section of KSU who encouraged us in our work.

All the figures in the book were drawn with the help of graduate students. DC would like to personally thank Dr Chandrasekhar Azad Kashyap and Dr Hemant Kumar Singh who tirelessly rendered support in accomplishing this task.

DC had the full support of his family members, especially his wife Dr Rama Chandrasekhar, who gave full moral support to complete this huge task in record time.

The authors thank Dr John Lund and Prof. Ranjith Pathegama who made initial reviews of the book's content and made several useful suggestions.

We thank the Director of the Indian Institute of Technology (IIT) Bombay and the Rector of KSU for their support during the writing of this book.

Chapter 1

Introduction

There is reason to be optimistic about geothermal energy. The exciting period is beginning where anomalous sources of heat are treated as systems. To develop geothermal energy as an important resource one must identify anomalous thermal sources and understand their genesis and geometry.

Grant Heiken, Los Alamos National Laboratory

Under the 6D scenario, following the business as usual trend, oil remains the most important primary energy source. Another three decades from now, under this scenario, global energy demand will grow by 70% and CO_2 emissions will grow by 60%. This trend is disastrous for the economies and future generations will be forced to live in an unpleasant environment. Curtailing the use of oil and supplementing it with renewable energy sources may improve the situation, but unabated use of old coal based thermal power plants and failure to implement clean coal technology based power generation systems will undermine this situation. Efforts made by all the countries to counter the effects of the 6D scenario will be fruitless if control over coal is not exercised. Electricity being the primary demand for all economies, a system that can substitute this primary energy should be able to generate baseload power with 97% efficiency and a minimal lay-off period while simultaneously reducing CO_2 emissions. Countries under the Intergovernmental Panel on Climate Change (IPCC) convention of Parties (COP) meetings are not able to pledge to reduce use of fossil fuels primarily because of the underdevelopment of a robust electricity generating renewable energy source that has zero carbon foot print. Technological advancement made over the last decade is giving hope to achieve the 2D scenario by 2030. This hope will become reality only when dependence on fossil fuels shows a declining trend. According to the Energy Sector Carbon Intensity Index (ESCII), if the business as usual scenario continues, making 6Ds as the expected emissions in 2050, then the world is calling for a disaster (Figure 1.1). According to the Energy Technology Perspective 2014 (IEA, 2014e), electricity generation through renewable energy sources is on track and making convincing progress. Thus, there seems to be some hope to bend the CO_2 emissions curve by 2050 (Figure 1.1). Between 2006 and 2013 electricity generation grew by 5.5% and was expected to reach 40% in another five years. This means the renewables will be able to generate about 6850 TWh by then.

Generation costs are posing a challenge to meet 2D's from renewable energy sources. Although renewable energy sources, other than geothermal, are being

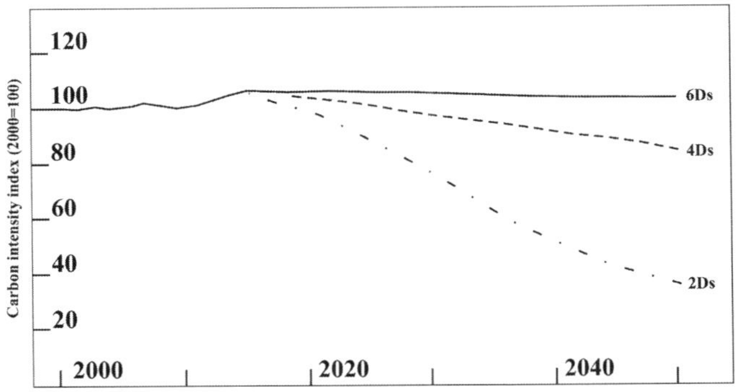

Figure 1.1 Carbon intensity index showing the influence of renewables in the three scenarios (6D: six degrees, 4D: four degrees, 2D: two degrees; adapted from IEA, 2014e).

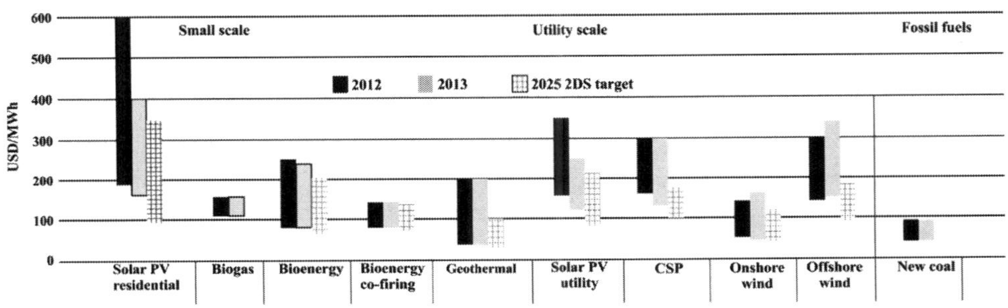

Figure 1.2 Levelised cost of electricity from renewable energy sources (IEA, 2014e).

discussed at length, given wider incentives, encouraging cost reduction in manufacturing the necessary components, they are not able to make substantial progress when levelised costs are considered. The levelised cost of electricity from various renewable energy resources published by IEA (2014e) is shown in Figure 1.2.

Although solar photovoltaics (PV) and Concentrated Solar Power (CSP) are reported to be surging ahead of other renewables, with double digit growth and pushing the global share of renewables to 20% (IEA, 2014e), the levelised cost either for solar PV (residential and utility) or CSP is not able to compete with geothermal energy. In fact, new coal technology is still able to be the leader in 2050 and hence coal will rule the electricity domain even after two decades. Perhaps this may be due to the large land and water requirements for solar PV compared to geothermal energy sources. Geothermal power plants need 1 acre/MWe (4047 m^2) while solar PV and wind power need 7 and 2 acre/MWe (Chandrasekharam et al., 2014a). Once the enhanced geothermal systems (EGS) technology matures, the geothermal energy source will be a zero carbon source, and with CO_2 being used as the circulating fluid to extract the heat, this technology will have a double advantage of controlling CO_2 emissions and generating electricity as well. This may even help coal based thermal plants to continue their operation without counteracting the emissions progress under the 2D scenario. Among all the renewable energy sources, geothermal energy alone can supply baseload electricity and does not require any backup support unlike other renewables. To meet the 2D target (Figure 1.1), countries have to work hard to bring down the CO_2 emissions per unit of electricity by 90% by the targeted date. This target reduction is possible by using geothermal energy sources as the primary source in the energy mix because the CO_2 emissions by geothermal energy power plants is only 0.893 kg CO_2/MWh while oil based power plants emit 817 kg CO_2/MWh and gas based power plants emit 193 kg CO_2/MWh (Chandrasekharam and Bundschuh, 2008). The countries that have easily accessible geothermal energy source like Eritrea, Djibouti and Yemen can adopt policies to include this as a primary energy mix for electricity generation. These countries heavily depend on imported fossil fuels to support electricity generation and transport, thereby increasing their energy security risk as well as their food security risk due to the supply volatility of oil. Those high temperature geothermal systems lying untapped, as discussed in this book, if developed, can make a difference in the socio-economic growth of the country in terms of access to cheap electricity, water, and also help the country in reducing CO_2 emissions and earn carbon credits. Saudi Arabia on the contrary, can offset part of its fossil fuel use for electricity generation by using its geothermal energy sources and reduce CO_2 emissions. It can also save additional CO_2 by using geothermal energy sources for its desalination processes and provide fresh water for the growing population in the future, for drinking and agricultural purposes. The emissions reduction strategies of these countries have been adequately deliberated in this book. Thus both oil exporting as well as oil importing countries have the opportunity to be energy independent and reduce food imports and become food secure countries. In all the countries discussed in this book, except Saudi Arabia, a large population lives in rural areas that have no proper infrastructure to improve its socio-economic status. For example, the urban population in Djibouti has access to electricity and water. Those living in remote places like in Lake Abhe Bad, have no access to electricity and live in poverty, while this site has plenty of geothermal energy sources that can support millions around the lake.

All the countries around the Red Sea, except Eritrea and Egypt, receive very low rainfall and lack proper surface drainage systems. As a result, these countries are stressed for water and affluent countries, like Saudi Arabia, desalinate the sea water to meet fresh water demand for drinking and for agricultural purposes. To date Saudi Arabia utilises 17 million kWh of electricity to meet a 235 L/day fresh water demand for its 29 million population. At present about 33 desalination plants are in operation in Saudi Arabia (Chandrasekharam et al., 2014a). With a 6% annual population growth, by 2030 the demand for the additional electricity needed to generate fresh water will increase manifold. This will has an effect on CO_2 emissions. These countries around the Red Sea are fortunate to have active volcanic and rift systems and an active and young spreading centre in their neighbourhood that can supply heat for decades. While the world is debating on the reduction of emissions by 2030 and 2050, these time spans, on a geological time scale, are fractions. These countries can evolve better systems of living with improved socio-economic and carbon free environments for centuries. As discussed in this book, the hydrothermal systems have evolved since 31 Ma while the high heat generating granites have evolved since 900 Ma. These countries can set an example as leaders in the world in evolving carbon mitigation strategies by using geothermal energy sources.

The countries around the Red Sea experience extreme weather conditions, where, especially in summer the temperatures reach >40°C and hence a huge amount of electricity is consumed to support air conditioning and ventilation systems. More than 70% of electricity is consumed for space cooling by Saudi Arabia (WEO, 2012). Use of fuel oil and direct combustion of crude oil to generate electricity to support space cooling has resulted in extremely low average power plant efficiency and excess emissions of CO_2 (WEO, 2012). Because of subsidies given for electricity, individual incentives to deploy more energy efficient systems have eroded there, resulting in huge energy wastage. Although the initial capital cost of geothermal energy is higher than other renewable energy sources, in the long term, these costs are absorbed since the systems work for many years, supplying baseload power with least down time and low maintenance costs (Figure 1.2). In addition, the fuel's cost, unlike the fossil fuels, is zero.

Besides electricity, the countries around the Red Sea face severe problem of fresh water shortage to support agricultural activity. For example, Yemen experiences increasing aridity due to climate change and the demand for water for agriculture and urban needs is ever increasing. Like energy, the water requirements will grow manifold in the future. Countries located in semi-arid zones are vulnerable to four types of water scarcity, of which two are natural and two are induced. This scarcity, when fuelled by intense population growth, escalates and manifests itself in socio-economic collapse (Falkenmark, 1989). However, solutions to such problems are at hand and it is up to an individual country to realise the solution and execute it. Yemen imports 100% of its food and the country's staple diet is wheat and rice. Because of this, the country faces food insecurity and child and maternal malnutrition is highest in this country compared to other countries in the world (WFP, 2010). In addition to this, the country has a high population growth (>3%/year), high poverty rate (35%) and poor infrastructure. Nearly 96% of the population are net buyers of food and are susceptible to fluctuations in the global food market. The situation is similar in Djibouti. In 2005, due to a worldwide increase in commodity prices, the purchasing power of individuals in Djibouti decreased drastically, gripping the country in poverty. Although the report (WFP, 2010) states that the country lacks natural resources, it is the inability of the government to develop its huge geothermal resources and the reluctance of the donors to help the country to realise its natural resources which have downgraded the country's ability to come out of the poverty and become energy and food secure (WB, 2012). Geothermal energy sources are plenty available around Lake Asal and Lake Abhe Bad, where exploratory and pilot power plant works were conducted and proved the capacity of the resources. 1 MWe (8 million kWh) of electricity can support about 6000 people (Vimmerstedt, 2002, Chandrasekharam and Bundschuh, 2008), and the Djibouti geothermal sources have the capacity to generate more than this quantity. Had the amount of financial aid the country obtained been utilised for developing these natural resources, the country could have become energy independent and food secure.

The situation in Saudi Arabia is different. The country has abundant oil and gas resources and controls the world economy. However, its inability to control excess utilisation of fossil fuels, not adopting energy efficiency systems, and not developing

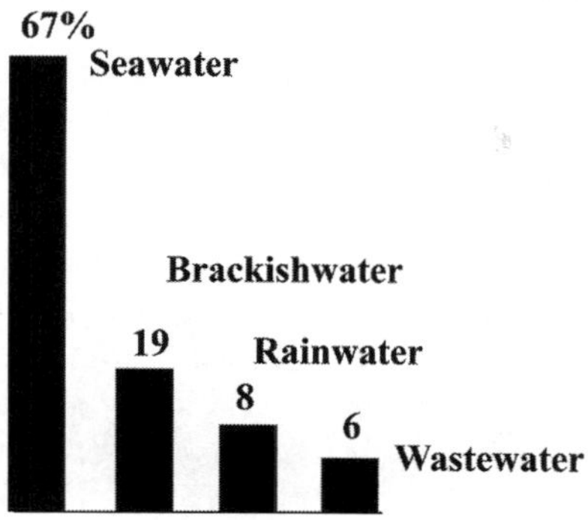

Figure 1.3 Feed water in desalination plants (adapted from Mezher et al., 2011).

its geothermal energy sources, is causing drastic climate related problems in terms of flash floods and a continuous increase in air temperature due to excess CO_2 emissions. The country's staple diet is wheat and barley that are grown in the west coast through irrigation. Small check dams and shallow groundwater resources support agricultural activity in this region. Domestic water supply is supported through desalination plants. The fresh water demand in Saudi Arabia is about 19 billion cubic metres (BCM) with 83% of the demand being from the agricultural sector. The country produces about 2 BCM of desalinated water annually from fossil fuel based processing plants (Chowdhury and Al Zahrani, 2015). With growing demand from the domestic and industrial sectors, restrictions are imposed on the agricultural sector to decrease domestic production and increase imports of staple food. From 2016 onwards, the country will adopt a policy of importing the entire domestic demand of wheat and barley and phasing out domestic production of these food items. The fresh water demand for agriculture can successfully be accomplished by generating fresh water through a desalination process using the currently available hydrothermal resources in Al Lith and Jizan (Lashin et al., 2014, Chandrasekharam et al., 2015b). In addition to fresh water, additional electricity can also be generated through this process. The feed water in desalination plants need not only be sea water (Figure 1.3). However, agricultural return water, wastewater, and brackish water as feed water for desalination plants may reduce the cost of desalination and reduce the power consumption.

In addition to the domestic and agriculture sectors, water is also required for energy production. Over 583 BCM of water were consumed by the energy sector in 2010 and by 2030 consumption will increase by 85% (IEA, 2012). This demand is apparently linked to the growth in population and the need to increase GDP growth through industrial activities (IEA, 2012). The fossil fuel and nuclear powered plants consume significant amounts of water. Water is also required to irrigate crops to support biofuels based power plants. Further, solar PV panels use water for cleaning to maintain production efficiency in countries like Saudi Arabia (Segar, 2014). Solar PV desalination plants operate at 20% efficiency and can generate 5000 cm^3/day of fresh water (Ahmad and Ramana, 2014). The water requirement of geothermal power plants is low and these power plants can generate fresh water for their own consumption.

Hence the above countries face two challenges: 1) to have a guaranteed electricity supply to meet ever growing demand and 2) to reduce CO_2 emissions. Guaranteed electricity will ensure water and food security and CO_2 emissions reduction will ensure sustainable development. No doubt, there are other options too that can meet these two challenges. Nuclear energy is one option that will meet both the challenges, but increasing global concern is a hindrance to its future accelerated growth. Hydropower development is beset with environmental concerns. Changes in weather patterns (poor rainfall) due to global warming is causing severe setbacks to already existing dams. Solar and wind power are other options, but they cannot supply baseload electricity and solar PV, as mentioned earlier, they need a large land area, while both solar PV and wind need back up power support. Biomass is yet another option, but biomass emits black carbon which is more harmful than carbon dioxide. The rural populations in Ethiopia, Djibouti, Yemen, and Eritrea depend on biomass to support their energy requirements. The black carbon affects the people around the source causing severe health related issues. However, this source was never given serious consideration. Thus, biomass is not a comfortable renewable energy source. The question is, what should be the criteria for considering renewable energy as the future energy, to meet growing demand and meet the 2D scenario? The following are the criteria: a) the source should be large enough to support long lasting energy supply to generate electricity and meet the present and future demand, b) the source should be easily accessible and economically viable, c) the source should be available over a large geographical boundary, d) the source should have the least carbon foot print and e) the source should be able to support baseload electricity supply without any backup support. Geothermal energy meets all the above criteria. Now that EGS technology is maturing, electricity should be available in the backyard of each community. In spite of the availability of substantially large geothermal energy sources around the Red Sea, some of the countries around it (Eritrea, Ethiopia, Djibouti, and Yemen) are still reeling in poverty and some of the countries are not able to control CO_2 emissions (Saudi Arabia and Egypt) or generate fresh water to support communities and food production.

This book provides sufficiently large information related to geothermal resources in the countries around the Red Sea, and assesses their potential and advantages for using this energy both by poverty ridden countries and countries that need water and food security for their future generations. Geothermal exploration methods are also included for those who wish to understand the fundamentals related to exploration techniques.

EGS technology and its potential in countries around the Red Sea that have the shield remnants are described adequately.

Chapter 2

Electricity demand and energy sources

Let us remember that oil and gas have not replaced coal, but grown with it. Other resources will come along. We will need all energy sources and give priority to those that emit little or no CO_2.

Speech given at the World Energy Council Executive Assembly 2006, Tallinn, Estonia, 6 September, by WEC Chair-Elect Pierre Gadonneix

2.1 WORLD OVERVIEW

Demand for energy is growing at a pace that the countries across the world are not able to keep up with. Since energy is closely associated with the socio-economic growth and gross domestic product (GDP) of any country, wars are being fought to have control over energy sources. Figure 2.1 shows the anticipated primary energy demand of countries around the world.

In Table 2.1 certain contrasting patterns emerge. China appears to be on the top of the energy ladder with a demand of 47,217 million MWh for a population of 1,387 million in 2035. In contrast to China, population and energy demand do not show a strong relationship in the Middle East. Here, the population is projected to grow from 199 million (2010) to 293 million in 2035, but the energy demand is far above that of South East Asia. Perhaps this is due to the rate of urbanisation. The Middle East has a very high urbanisation rate (74%) compared to India (43%). Urban living demands higher comfort levels, and hence requires more energy. For example, urbanised regions require continuous space cooling and other continuous energy services that consume considerable amounts of energy. Besides urbanisation, the Gulf Cooperation Council (GCC) will be a future trading centre with a projected economy of US$ 2 trillion (The Economist, 2009). This means that several oil- and gas based industries will be hosted by these countries and these countries will be strong trading partners with other economies.

The global energy scenario today is different from the one that existed about a decade ago. The main thrust is to maintain sustainable development and reduce CO_2 emissions without much sacrifice to national GDP growth and poverty removal. Several new policies have been evolved by several countries to achieve this goal. These policies include targeting renewable energy resources; methods to improve energy efficiency and save electricity; and methods to reduce greenhouse gas emissions. The policies also include decommissioning inefficient, old fossil fuel based power plants. However, these

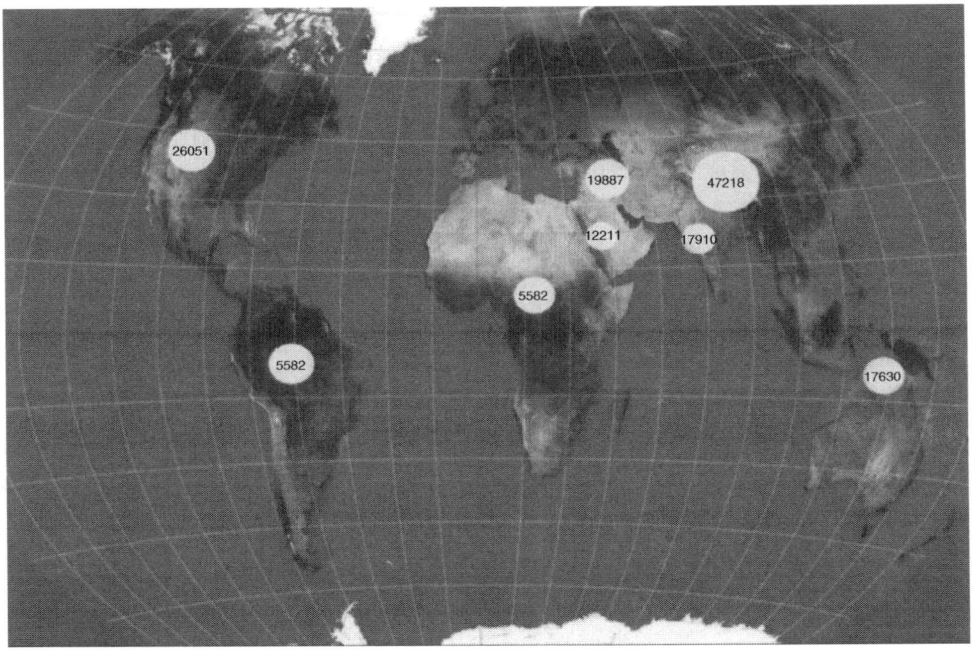

Figure 2.1 Predicted primary energy demand in million MWh in 2035 (Source: International Energy Agency, 2012).

Table 2.1 Predicted primary energy demand and population in 2035 (source: International Energy Agency, 2012).

Country/region	*Million MWe*	*Millions*
China	47217.8	1387
USA	26051.2	377
S America	5582.4	782
Europe	19887.3	599
Eurasia	15933.1	464
Africa	11978.9	1730
Middle East	12211.5	293
India	17910.2	1511
Japan	5117.2	118
SE Asia	11630	4271

policies are not action based and hence at the United Nations Framework Convention on Climate Change (UNFCCC) Cancun meeting (450 Scenario), it was decided to limit the increase in the rise of global atmospheric temperature to 2°C by controlling the CO_2 concentration in the atmosphere to 450 ppm.

Several countries are limping back to recovery after the effect of global economic recession in 2009. The worst affected are countries that do not belong to the Organization for Economic Cooperation and Development (OECD). Energy demand is very

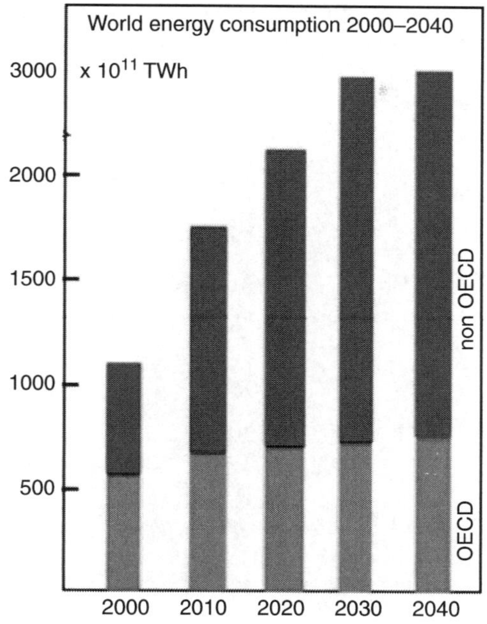

Figure 2.2 Historical and predicted world energy demand (source: International Energy Agency, 2013).

high in non-OECD countries compared to OECD countries. This is due to emerging economic situations in the former compared to the latter, where the economy is relatively stable in OECD countries. The energy demand is about $3,000 \times 10^{11}$ TWh per year in developing countries compared to the OECD countries where it is about 600×10^{11} TWh per year (Figure 2.2, International Energy Agency (IEA) 2013).

The fossil fuels will almost certainly rule the energy sector at least until 2040 (Figure 2.3). Therefore, electricity consumption and transportation will remain the main contributors to CO_2 emissions until any of the renewable sources can generate baseload electricity and contribute large enough quantities of electricity into the energy mix for developing economies. Only geothermal energy meets these conditions and once the enhanced geothermal system matures, all the countries can vigorously implement energy policy to drastically reduce CO_2 emissions and contribute to controlling global climate change. Therefore, countries should focus on developing enhanced geothermal systems (EGS) technology to tap the earth's heat source and implement policies to reduce greenhouse gas emissions.

Countries with oil and gas resources and the countries without them face the same problem. A good example is the political unrest in the oil rich Middle Eastern countries and those countries without the oil resources that include many OECD and non-OECD countries. Those countries with fossil fuel resources are facing political problems and are not able to stabilise their economy through oil and gas exports due to sanctions, and those countries that depend on imported fuels are facing similar socio-economic problems due to an increased demand for energy sources. Thus, energy situation may or may not improve in the future, and those countries depending on fossil fuels imports

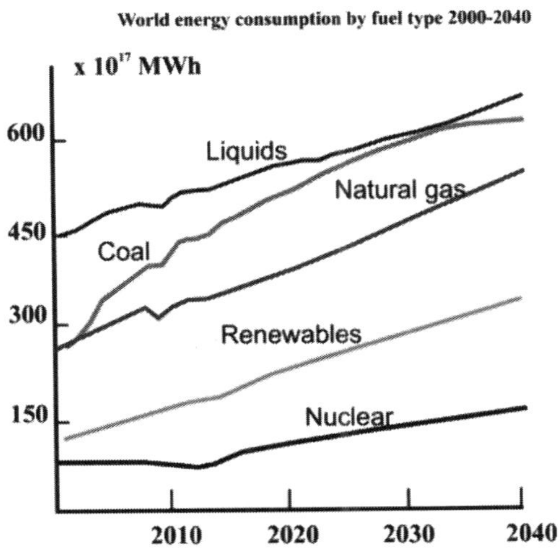

Figure 2.3 Historical and predicted world energy consumption by fuel type (IEA, 2013).

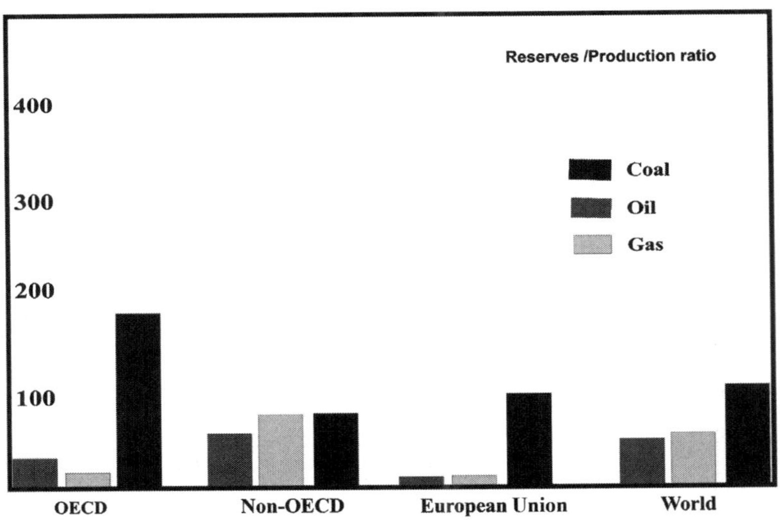

Figure 2.4 Recent reserves-to-production ratio, fossil fuels in different sectors in the world (source: BP, 2013).

for their economic growth should think seriously of methods to be energy-independent. Here, geothermal energy will play a crucial role in the future, not only to bridge the demand–supply gap, but to make countries free from the influence of external factors for their growth.

Countries that have fossil fuel resources have to do a balancing act to maintain a certain reserve-to-production ratio to maintain economic growth. This is especially true with the non-OECD countries that have small reserves. Figure 2.4 gives an idea of recent reserves-to-production ratio of OECD and non-OECD countries (BP, 2013).

2.2 REGIONAL ELECTRICITY MARKETS, SOURCE MIX AND FORECASTS UNTIL 2030

2.2.1 Egypt

Amongst all the African countries, Egypt – with a population of 90 million – is a non-member of OPEC (the Organization of the Petroleum Exporting Countries), and is the largest oil-producing country in that continent. It is also the second largest country in gas production in Africa. Besides oil, the country also levies fees on all the vessels – including oil tankers – passing through the Suez Canal, thus earning an additional revenue from oil. The country's energy needs are met by oil and gas. Oil based power plants generate 25 TWh of electricity and gas based power plants generate 117 TWh of electricity per year (Table 2.2), emitting 204,777 kilotonnes of CO_2. The per capita CO_2 emissions are 2.63 tonnes. Thus, Egypt is a large consumer of electricity when compared to other states bordering the Red Sea, such as Eritrea, Ethiopia and Djibouti. Other renewable sources like biomass, solar photo voltaic (PV), wind and hydropower also support the generation of electricity in Egypt, taking the total electricity production to 157 TWh (US Energy Information Administration (EIA), 2013).

The GDP growth of Egypt is drastically affected by political unrest and stood at about 6% in 2013 compared to Ethiopia (7.8%) and Djibouti (6.3%) (EIA, 2013).

Egypt produces oil from small, interconnected fields located in the Gulf of Suez, the Nile Delta, the Western and Eastern Deserts, Sinai, and the Mediterranean Sea. These fields are small and limited and hence the production has drastically declined from 900,000 bbl/day in 1990 to 720,000 bbl/day in 2012 (EIA, 2013). Ever-growing demand for oil – at an average rate of 3% per year – and declining production from the country's own fields, plus subsidies imposed by the government is affecting the country's economy. Egypt's diesel consumption is greater than its gas consumption, forcing the country to import diesel (EIA, 2013).

The energy sector is an important part of the economy for Egypt. The country's economy is energy intensive. Tourism and manufacturing, which are both energy intensive, are two important components of its economy. During the last decade these two components represented about 25% of GDP. Although Egypt produces oil and gas, a major part of these are being consumed within the country and only a small percentage is exported. Hence, it is more important for the country to develop renewable energy resources, rather than increasing the import of diesel to meet the growing demand.

As of today, Egypt generates 147 TWh of electricity, most of which is generated from oil based power plants. Of that, 16 TWh are lost during transmission and the remaining is consumed. Gas based power plants generate nearly 75% of the electricity while only 14% is generated from oil based power plants. The hydroelectric source generates about 10%, while the remaining is produced from wind, solar, and biomass. Egypt's estimated gas reserves are only 78 trillion cubic feet as of 2012, and new discoveries are being made towards the northern coast of the country, in the Mediterranean. However, considering the growing demand, the production may not be able to keep up with the demand since the reservoirs have shown a decline in production since 2008 (Figure 2.5).

Due to the rapid growth of Egypt's economy, the primary energy demand has grown to 4.5% in the last couple of decades. Because of this demand, the country's

Table 2.2 Population, CO_2 emissions and electricity consumption by countries (excluding Sudan) around the Red Sea.

	Population (million)	*P density per sq. km*	*2000 CO_2 kt*	*2012/13 CO_2 kt*	*CO_2 from gas kt*	*CO_2 from oil kt*	*CO_2 tonnes/person*	*Energy use kt oil eq*	*Electricity $\times 10^9$ kWh*	*Electricity from oil $\times 10^9$ kWh*	*Electricity from gas $\times 10^9$ kWh*	*Power consumption $\times 10^9$ kWr*	*Power per capita kWr*
Egypt	82.0	82.4	141326.2	204776.3	87494.6	89181.4	2.6	88208.5	157	25	117	139	1742.9
Eritrea	6.1	62.7	608.7	513.4	0.0	491.4	0.1	598.8	0.33	0.33	0	0.3	48.7
Ethiopia	91.7	94.1	5830.5	7900	0.0	5148.5	0.1	32114.0	5.2	0.03	0	4.7	52.0
Djibouti	0.9	37.7	403.4	539.0	NA	539.0	0.6	NA	0.73	0.73	NA	0.26	336.0
Republic of Yemen	23.9	45.2	14638.7	21851.7	1679.5	18676.0	1.0	7260.5	6.3	4.9	1.3	4.6	193.5
Saudi Arabia	28.8	13.4	296935.3	464480.6	152330.8	291053.5	17.0	197070.1	251*	66	108	227	8161.2

*includes other sources

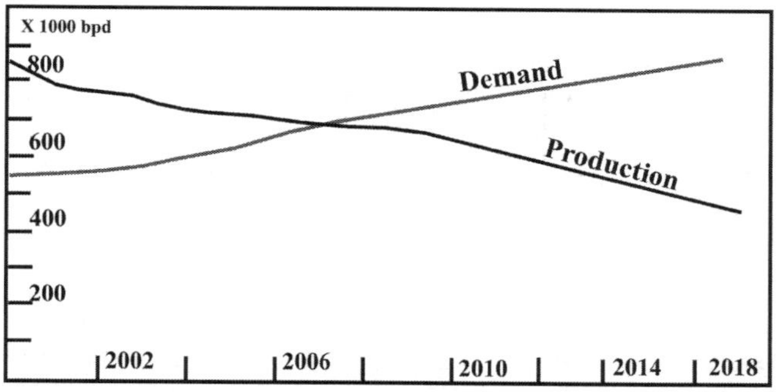

Figure 2.5 Oil production and consumption, Egypt. Data beyond 2010 are projections (source: adapted from EIA, 2013).

economy is dubbed a carbon intensive economy. The manufacturing industry consumes nearly 11% of gas, followed by fertilisers and cement, consuming about 10% and 8%, respectively, and the rest is consumed by power generation plants. This growth has been a cause for concern due to declining oil production (Figure 2.5) since 2008. Hence, there is a strong reason for the country to look for other energy sources to sustain its supply of energy for a long time.

Egypt has keen interest in developing nuclear energy to support its ever-growing demand for power. Between 1954 and 1961 Egypt built up sufficiently skilled human resources dedicated to handle nuclear power, and a power project was established in 1962. However, it was not commissioned due to the conflict with Israel between 1967 and 1973. In 1974 this interest was revived and the country signed a cooperative agreement with the USA for a 900 MWe reactor, which produces electricity from pressurised water reactors. This was again stalled due to the Chernobyl accident in 1986. An attempt to revive the programme was again stalled by the recent Fukushima nuclear disaster in Japan in 2011. Thus, from 1962 to 2011 the energy situation has not really progressed in the country. The IEA in 2011 brought out a road map to develop the nuclear power industry in Egypt.

2.2.2 Eritrea

Eritrea, with a population of 6.1 million (Table 2.2), made satisfactory progress after independence in 1991 in its socio-economic structure, with economic growth of about 11% in 1997 (World Bank, 2012). It continued growing, to become one of the fastest growing economies of all the African countries in 2011. Its current GDP growth stands at 14%. Only 32% of population has access to electricity and the rest depend on other energy sources. The country's economy is supported by agriculture, mining, and direct foreign investment. However, the country itself is poor, with an average annual per capita income of US$ 403. About four million people live in rural areas with economic support from rain-fed agriculture. Frequent droughts restrict the food supply to only 70% of the population and this is pushing the rural population below the poverty line. Thus, the country's setbacks are food security and electricity supply. By the end of 2012 the World Bank is optimistic that half of the population can be lifted above poverty status under the Millennium Development Goal (MDG) (World Bank, 2012). Life in urban areas is better than the life in rural areas because of the accesses to electricity. The rural population still depends on traditional biomass fuel to support their energy requirements. Currently, the country generates about 338×10^6 kWh from oil based power plants. The per capita electricity consumption is about 49 kWh (IEA, 2013). The total CO_2 emissions are 514 kilotonnes, with oil contributing 492 kilotonnes only. The rest comes from biomass sources. Interestingly, the CO_2 emissions in 2000 were much higher (609 kilotonnes, see Table 2.2) compared to 2013. While the 2013 per capita electricity consumption by Eritrea is closer to Ethiopia's 2013 figure (51.96 kWh) than to any of the other four countries under consideration, the per capita emissions of CO_2 by Eritrea are higher (0.09 tonne) compared to Ethiopia (0.07 tonne). Although sites for developing hydropower were identified on the rivers Tekezé and Anseba, with a combined estimated capacity of about 23,000 GWh per year, they are yet to be developed due to lack of infrastructure (Bartle, 2002). Eritrea imports nearly 5,000 bpd of oil and refined oil products.

The Renewable Energy Centre (REC) in Eritrea is promoting different technologies to reduce CO_2 emission from biomass energy sources from rural areas, and is able to save about 300,000 tonnes/year of carbon dioxide emissions. Other rural technologies that are being adopted to control carbon dioxide emissions include a fuel-efficient charcoal stove, and a fish cooking stove for the coastal fishing population. Data on black carbon (BC) emissions from biomass energy sources is not available. Black carbon emissions are more harmful to the rural population than CO_2 and are detrimental globally to ice caps – including those over the volcanoes in East Africa (e.g. Mt. Kilimanjaro) – as the particles darken the ice and absorb more sunlight. Since the life of BC in the atmosphere is only about a week, whereas CO_2 stays in the atmosphere for at least a century, controlling BC emissions through non-polluting energy sources like geothermal, wind, and solar is very important.

Geothermal sites occur around active volcanoes like Alid, south of Massawa, and are waiting to be developed for power generation and direct applications like timber drying, greenhouse cultivation space, cooling, and heating. The advantages of geothermal energy are discussed in the forthcoming chapters.

2.2.3 Ethiopia

Ethiopia, with a population of 92 million (Table 2.2), is one of the poorest countries in the sub-Saharan region. In the recent past, Millennium Development Goals (MDG) of the World Bank were successful in setting the country's GDP is on the road to recovery. From US$ 452 per capita in 2000, it is expected to grow to US$ 698 by 2015, thus reducing those living below the poverty line to below 30% (World Bank, 2012). The Ethiopian government is also very optimistic that it can bring substantial changes in the country's socio-economic status by implementing several programmes of the World Bank, such as increasing the economic growth to 11% per year and reducing the poverty level from 30% to 22% by 2015. Although GDP growth is high (8%) relative to other countries considered here, and the economy is showing signs of improvement, the per capita income is the lowest of the countries considered here, compared to the world average (Ethiopia's per capita income is 5.7% of the world mean, World

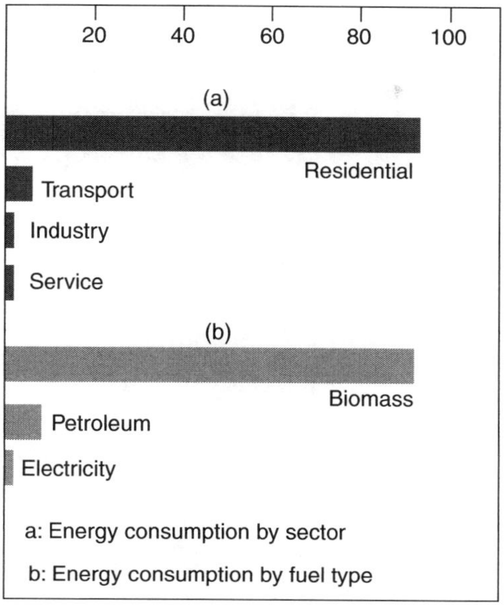

Figure 2.6 Total energy supplied from all sources for Ethiopia (source: IRENA, 2014).

Bank, 2008). Ethiopia's economic growth depends on its energy availability. The delay in constructing a dam across the Nile for a huge hydroelectric project is one of the major hurdles in growth development. The advantage Ethiopia has over Egypt is the availability of substantial quantity of geothermal energy in the rift valley. It is essential for Ethiopia to develop this energy source and become energy-independent.

Ethiopia's electricity generation capacity has grown to 2,177 MWe in 2014 from 2,000 MWe in 2009 (Federal Democratic Republic of Ethiopia (FDRE), 2014), a meagre 177 MWe addition in five years. Electricity supply showed a remarkable increase from a 41% coverage of the country in 2010 to 53.5% in 2013. Most of this development has come from hydropower projects. Under the Ethiopian government's Growth and Transformation Plan (GTP) its electricity generation capacity is expected to touch 10,000 MWe and the supply grid to cover nearly 75% of the rural areas by 2015 (FDRE, 2014). Implementation of this plan requires over US$ 2 billion/year. The main barrier in executing this plan is the Ethiopian Electric Power Corporation, the key agency for power generation and supply, which does not have sufficient infrastructure to handle such mega projects (http://www.reeep.org/publications).

For the Ethiopian population, biomass is the main fuel source, providing nearly 92% of the energy supply. Oil and gas are the next sources, accounting for 7% of the energy supply (Figure 2.6).

The residential sector consumes about 96% of energy, followed by transport. The countries poor economic growth and low per capita income is reflected in the energy consumption by the industry and service sectors (Figure 2.7). The energy sources and usage between rural and urban population varies drastically.

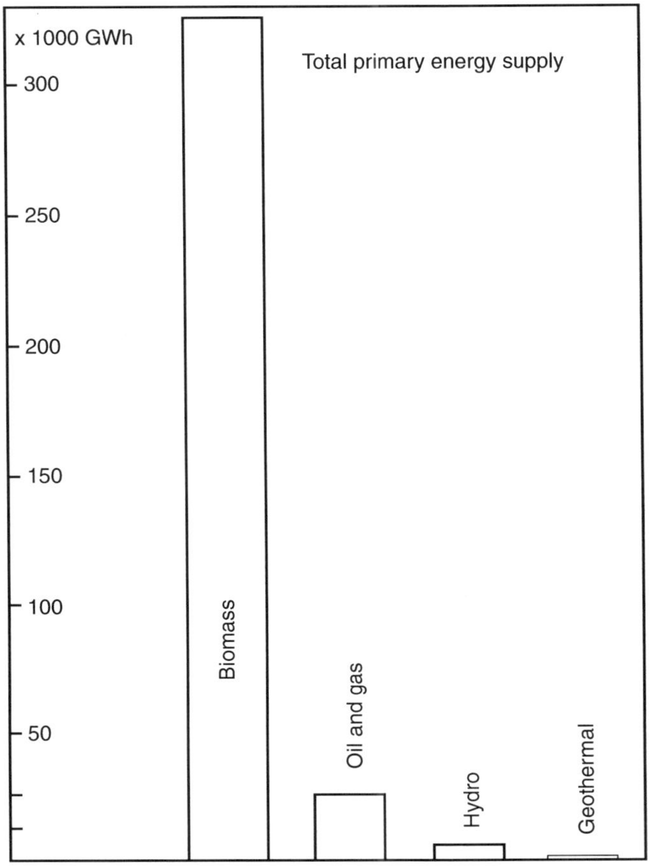

Figure 2.7 Ethiopia – energy consumption and fuel type (source: Tucho et al., 2014).

To generate about 10,000 MWe of electricity and supply 75% of the rural areas, Ethiopia needs large finance to support rural electricity demand (Figure 2.8). At the current rate, Ethiopia needs US$ 2 billion. For developing hydropower, which has been the backbone of electric power generation in Ethiopia all these years, the country needs at least 38% of the cost to be provided by additional financial support. It has about 18% available in the country, and needs at least 20% to be borrowed from external agencies (Ethiopia's Climate-Resilient Green Economy, FDRE 2011). This is a major challenge to the country.

2.2.4 Djibouti

Djibouti is the least populated country compared to other countries around the Red Sea, with a population of 0.9 million, and with a density of 37 persons in a kilometre (Table 2.2). The country's economy is supported by service activities because of its strategic location and free trade zone status. Food, energy, and other material to support the population are imported. The majority of population lives in Djibouti city.

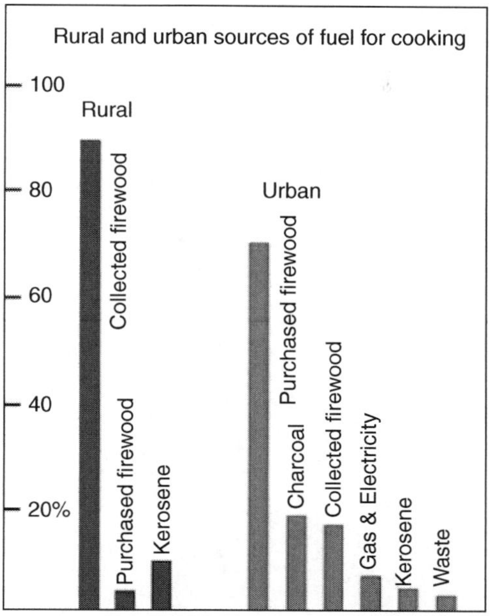

Figure 2.8 Ethiopia – energy consumption and fuel types between rural and urban population (%).

Due to lack of vegetation, poor rainfall, and poor agricultural activity, the population in the rural areas live in poverty, even though the country has a very rich geothermal resource. The annual GDP growth is about 4.5% with a per capita GDP of US$ 1,670 (World Bank, 2014). The country's electricity and transport are supported by 100% imported oil. Oil consumption has been more or less constant at 11 to 12 thousand bpd (Figure 2.9) except for a short period, between 2006 and 2008, when the country faced severe socio-economic problems related to the civil war, recession, and an influx of refugees.

The power and transport sector are supported by oil imports. The current installed power capacity is about 0.14 GWe (~0.9 billion kWh), while the consumption is only about 0.3 billion kWh (Figures 2.10 and 2.11). The per capita electricity consumption is 333 kWh, which is far less compared to its neighbouring countries around the Red Sea. However, Djibouti has huge geothermal energy resources at Lake Asal and Lake Abhe lying unutilised. This energy source, discussed in the succeeding chapters, if developed, would make Djibouti's economy surpass that of other neighbouring countries like Ethiopia, Eritrea, Egypt, and Yemen. Since the population of Djibouti is small, this energy could be transmitted to other, neighbouring countries.

The country's socio-economic problems between 2005 and 2008 appear to have had no influence on the electricity consumption (Figure 2.11).

On one hand, energy-starved non-OECD/developing countries, termed as emerging economies, are struggling hard to achieve sustained growth. Though these countries are attracting huge investments from rich countries, the setback these countries face is

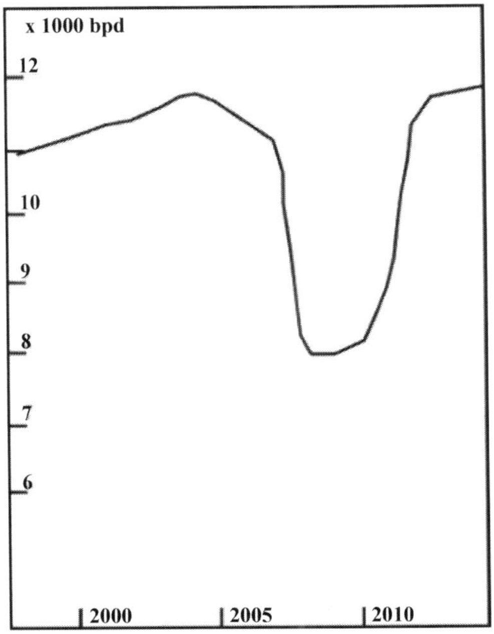

Figure 2.9 Oil imports and consumption, Djibouti (source: EIA, 2013).

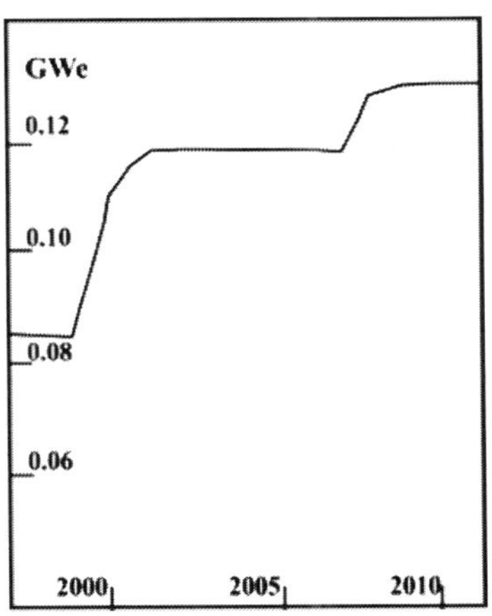

Figure 2.10 Djibouti – installed electricity (EIA, 2013).

a lack of sustained supply of electricity (energy) and fresh water. These energy-starved countries have huge geothermal resources waiting to be developed.

On the other hand, energy-rich countries (oil-rich countries), like the Gulf countries, face a severe shortage of drinking water and depend heavily on desalination. As shown in Figure 2.12, Eritrea, Ethiopia, Yemen, and Djibouti fall into the first category

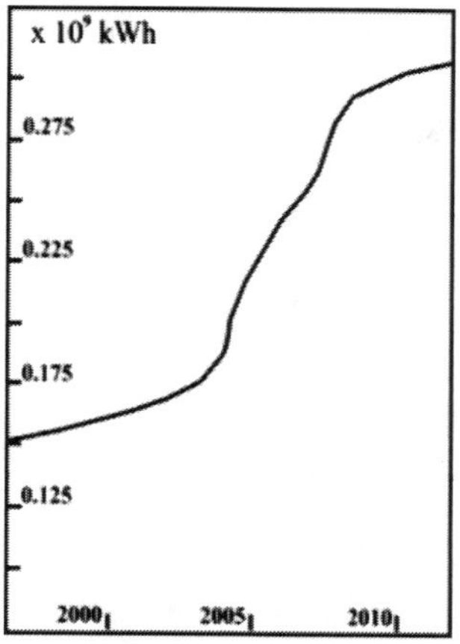

Figure 2.11 Djibouti – power consumption (EIA, 2013).

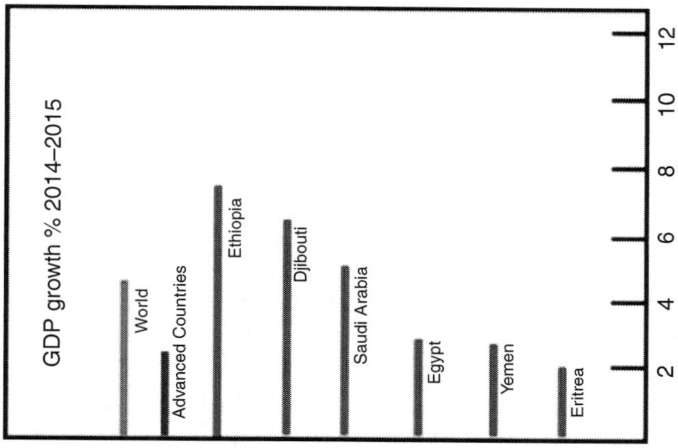

Figure 2.12 GDP growth of the countries around the Red Sea compared to the world and advanced countries (source: UNDP, 2014).

while Saudi Arabia and Egypt fall into the second category. Both Saudi Arabia and Egypt emit large volumes of CO_2 (Table 2.2) that are detrimental to them. The effect of high CO_2 emissions is already being experienced by countries like Saudi Arabia (Chandrasekharam et al., 2014a,b). Fresh water supply to its urban areas is also an issue in Djibouti. The urban population requires nearly 14 million m^3 of fresh water and the water is pumped using electricity at a rate of 436 Wh/m^3 water supplied. Besides electricity and transport, water supply also consumes a considerable amount of oil.

2.2.5 Republic of Yemen

The Republic of Yemen, situated towards the southern part of the Arabian Peninsula, is the poorest country in the Arab world, with a population of about 24 million and

Table 2.3 Oil and gas facts of Saudi Arabia vs OPEC (OPEC, 2014).

	Saudi Arabia	*OPEC*
OPEC statistical Bulletin 2014		
Population (millions)	29.9	439
Land area (1000 sq km)	2150	11862
GDP percapita US$	24847	8039
Value of exports (million US$)	377013	1581481
Valus of imports (million US$)	163900	879951
Value of petroleum exports (million US$)	321723	1112085
Proven crude oil reserves (million barrels)	265789	1206170
Natural gas reserves (billion cubic m)	8317	95034
Crude oil production (1000 b/day)	9637	31604
Oil demand (1000 b/day)	2994	9031
Crude oil exports (1000 b/day)	7571	24054
Natural gas export (million cubic m)	nil	224918

with an annual population growth of 2.4%. The current GDP is US$ 35.6 billion. Currently the country imports food worth US$ 721 million and fuel and energy worth US$ 416 million. The poverty level has increased to 55% of the total population in 2012, from 42% in 2009. Nearly 45% of the population have no food security and water resources are scarce (World Bank, 2014).

2.2.6 Saudi Arabia

Saudi Arabia, occupying the bulk of the Arabian Peninsula, is the third largest Arab country and the largest Middle Eastern country. Saudi Arabia has an estimated population of 29 million (World Bank, 2009) spread over an area of about 2.2 million square km. The country has the world's second largest oil reserves, is the world's largest oil exporter and is a member of OPEC (OPEC Annual Statistical Bulletin, 2014). Saudi Arabia exports 7.5 mbs/day of crude oil and 794,000 b/day of petroleum products. The status of oil and gas facts of Saudi Arabia compared to all the OPEC countries combined is shown in the following table, Table 2.3.

The export of oil and oil refined products accounts for 75% of Saudi Arabia's revenue. The country exports about 7.5 mbs/day of crude oil, nearly one-third of the export of crude from OPEC countries (Table 2.3). The country has proven oil reserves of 265 billion barrels which is nearly one-fifth of the reserves within the OPEC countries and produces 9.6 mbs/day crude oil (Table 2.3). Crude oil exports are over 7.5 mbs/d and the domestic use of oil is only about 2.1 mbs/d. However, most of the gas produced is consumed by the country to meet its energy demand, mainly for electricity generation and desalination.

Amongst the countries around the Red Sea under consideration in this study, the maximum per capita energy consumption is by Saudi Arabia, with 8,161 kWh per person (Table 2.2). With a growth in population of 6% per year, the per capita energy consumption is growing at 7.5% per year and the per capita consumption is likely to exceed 10,000 kWh by 2020 (Figure 2.13) (IEA, 2013, Chandrasekharam et al., 2014a).

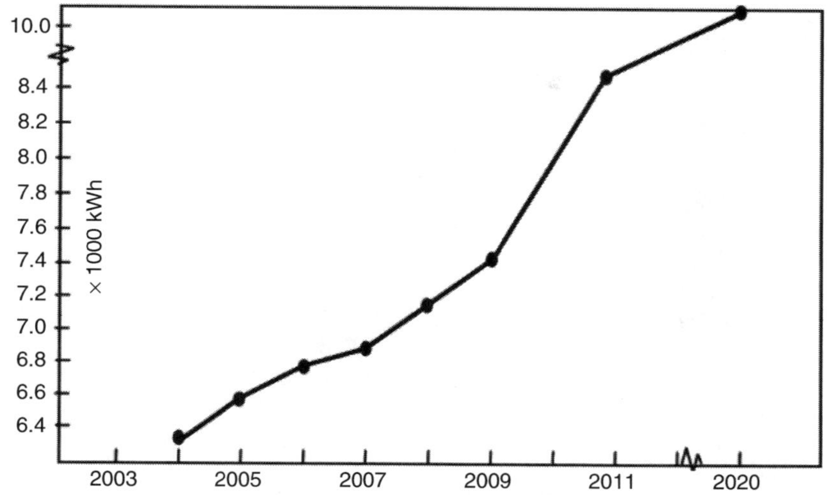

Figure 2.13 Saudi Arabia – per capita power consumption projections (Chandrasekharam et al., 2014a).

According to the EIA 2013 statistics, in 2011 the country generated 251 billion kWh of electricity: 66 billion kWh from oil, 108 billion kWh from gas, and the rest from other sources. This is generated by burning 500,000 barrels of oil a day (2011 figures). In summer, 900,000 barrels a day are needed to meet the demand. The current consumption of oil is 3.4 mbs/day (Chandrasekharam et al., 2014). According to projections, the country may need 8.3 mbs/day by 2020, with 3 million of those reserved for the power sector (Chandrasekharam et al., 2014, 2015). By the year 2020 the country's power generation capacity will reach 77 GWe from the current 44.5 GWe, with current population growth at the rate of 6% (World Bank, 2009).

Of late, Saudi Arabia is more inclined to export refined products rather than increase exports of crude oil. The major oil exploration and production company of Saudi Arabia, Saudi Aramco, is expanding the domestic refining capacity from 2.1 mbs/day in 2013 to 3.3 mbs/day by 2018. Thus the country will have a major global share in oil refining capacity: 5.7 mbs/day. Some of the major oil refinery units include the Aramco's refinery in the Persian Gulf, with a capacity of 400,000 b/day of heavy crude oil; the refinery at Yanbu on the Red Sea coast, with a capacity of 400,000 b/day; and a third refinery near Jizan on the Red Sea coast, with a capacity of 400,000 b/day. With the expansion of facilities, the total refining capacity of Saudi Arabia (Aramco) will be 3,344,000 b/day (Krane, 2015). The advantage of refining the crude oil instead of burning it for the generation of electricity is that valuable light and middle distillates like gasoline, diesel, and jet fuel can be separated from the crude which can be used for power production as well for export, thus increasing the revenue to the country (Krane, 2015). Predictions are that Saudi Arabia will maintain its crude oil production at the current rate and reduce the crude oil export by increasing the value added petroleum products to have better hold on its economy (Krane, 2015).

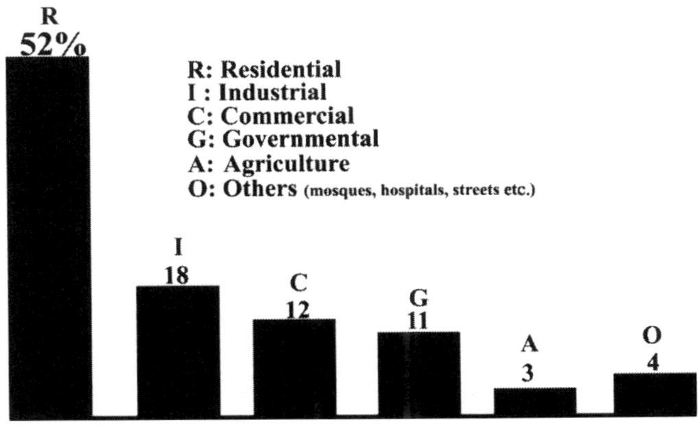

Figure 2.14 Electricity consumption by sectors, Saudi Arabia (adopted from Alaidroos and Krarti, 2015).

Nearly 80% of the electricity generated in Saudi Arabia is consumed in buildings. In Saudi Arabia, the residential sector is the greatest consumer of electricity, followed by industrial sector (Figure 2.14). The commercial establishments, which attract huge business activity and require constant electricity for space cooling have less demand compared to industrial establishments (Figure 2.14). This difference may be due to the fact that the industrial establishments function twenty-four hours a day, seven days a week, while the commercial establishments have set business hours. Further, the high consumption of electricity by industrial establishments is also due to desalination units that use considerable electricity. For example, 134 million kWh are consumed to generate 275 l/day of water from sea water (Chandrasekharam et al., 2014a,b, 2015). In future, the consumption of electricity by the residential sector will show tremendous increase due to 6% growth in the population. It is estimated that by 2020 Saudi Arabia will need 2.32 million new homes (Asif and Alrashed, 2012).

Thus, oil and gas play a crucial role in the economy and GDP of Saudi Arabia, which is a part of the Middle East and North Africa (MENA) countries. The MENA countries, having a large share in oil production compared to other countries, have a crucial role to play in controlling CO_2 emission and mitigating global climate change.

Beside oil and gas, Saudi Arabia is toying with the idea of using an energy mix to address growing energy demand, increase the proportion of oil it exports, and reduce greenhouse gas emissions by using renewable energy sources like solar photo voltaic, wind, biomass, nuclear, and geothermal energy (El Khashab and Al Ghamedi, 2015, Ramli et al., 2015, Ouda et al., 2015, Shaahid et al., 2013). In the case of the solar photo voltaic energy source, the advantage Saudi Arabia has is the number of sunshine days. The solar energy that Saudi Arabia receives varies from 4 to 7.5 $kWh/m^2/day$, which is far more than that received by Europe which varies from 1 to 1.7 $kWh/m^2/day$. The King Abdalla Centre for Atomic and Renewable Energy (KACARE), the key agency to promote renewable energy in the country, plans to enhance solar power installation from the current 0.003 GW to 41 GW by 2032. This includes two separate 25 GWe and 16 GWe installations (Nassif, 2012, El Khashab and

Al Ghamedi, 2015). Although these plans have been drawn up to save domestic oil consumption and increase oil exports – to increase country's revenue and decrease its own carbon dioxide emissions – Hashim Yamani, the President of the King Abdullah Centre for Atomic and renewable Energy, in a recent statement at the World Future Energy Summit, held at Abu Dhabi in January 2015, said that there will be delays in implementing solar projects by eight years due to the current decline in oil prices. Thus, wind biomass, and solar energy sources that are considered as cost-effective and environmentally-friendly energy sources and discussed in scientific literature since 1994 (Shaahid and Elhadidy, 1994, Eltamaly, 2013, Ramli et al., 2015), are not able to make significant contribution to energy mix as yet.

A royal decree was issued in April 2010 to create a renewable city as a mark of the country's commitment to develop alternative energy sources. Thus The King Abdullah City for Atomic and Renewable Energy (KACARE) was created in Riyadh, the capital of Saudi Arabia, which is the key agency to develop renewable energy resources and coordinate national and international energy policy. In 2011 KACARE announced that 16 nuclear power reactors will be constructed over the next 20 years with a budget of US$ 80 billion to meet 20% of Saudi Arabia's electricity demand. Generation of 17 GWe of electricity from nuclear power by 2032 was initially projected by KACARE, but now this target has been extended to 2040 (NWA 2015). The main reason for Saudi Arabia's interest in supporting nuclear energy for electricity generation and desalination is that the government has realised that the current dependence on fossil fuels to meet growing electricity and fresh water demand is unsustainable politically and economically. What the government had not realised in 2011 is the availability of huge quantities of heat source locked up in the rocks along the western shield region. Based on a cost analysis of the sort of time delay likely before Saudi Arabia can begin establishing nuclear power plants, a recent analysis concluded that nuclear power is a costly option for Saudi Arabia's power and water problem and the country will continue to use oil and gas to meet its future power and water demands (Ahmad and Ramana, 2014). What KACARE did not realise then, and do not realise now, is the country's geothermal potential. KACARE's website shows 1 GW of geothermal development by 2032. The basis for arriving at this figure is not known. In the following sections both hydrothermal and enhanced geothermal system (EGS) potentials are discussed in detail.

Chapter 3

Carbon dioxide emissions

The impact of energy decisions taken today will endure for the next century and beyond. All of us owe it to our children, to our grandchildren, to ourselves, and to our environment, to see this dialogue and cooperation through so as to lead to a successful and durable future.

Opening address by the Ambassador of US at the Baltic Regional Energy Forum June 12, 2007

3.1 WORLD OVERVIEW

Carbon dioxide concentration in the atmosphere has increased significantly during the last century compared to the pre-industrial years where the increase was steady. The concentration of CO_2 (carbon dioxide) in the atmosphere during the pre-industrial period was about 280 ppm, while the reported concentration of CO_2 in 2013 was 396 ppm which is 40% more than that recorded during the pre-industrial period (IEA, 2014b). This amounts to a mean increase of 2 ppm/year since the year 1800 (IEA, 2014b). The energy sector in the world is the major contributor of CO_2 because 82% of the world's primary energy comes from fossil fuels, which, when combusted, release greenhouse gases. Worldwide, the primary energy sector releases 69% of greenhouse gases (GHG), out of which 90% is CO_2, and the rest is made up of CH_4 (methane) and N_2O (nitrous oxide). CO_2 and black carbon (particulate matter) emissions are emitted at their highest levels from the combustion of biomass and incomplete combustion of fossil fuels and that accounts for 14% of total GHG (Figure 3.1). A large amount of CO_2 is emitted by coal-fired power plants compared to oil- and gas-fired power plants because they are more numerous than oil-fired plants, whose contribution to world electricity generation is only about 6%, and is expected to fall by about 4% in 2030 (Chandrasekharam and Bundschuh, 2008). Typically, coal-fired power plants emit 953 kg of CO_2/MWh, while oil-fired power plants emit 817 kg of CO_2/MWh, and gas-fired power plants emit 193 kg CO_2/MWh (UNFCC, 1997, Kasameyer, 1997, Chandrasekharam and Bundschuh, 2008). The CO_2 emissions projection indicate that the world is expected to face the worst climate related problems within another decade, when the CO_2 increase in the atmosphere will be steep if mitigation strategies are not adopted to control emission trends (Figure 3.2).

The CO_2 concentration in the atmosphere will soon likely to exceed the 500 ppm mark and, if the business-as-usual attitude of governments and people prevails,

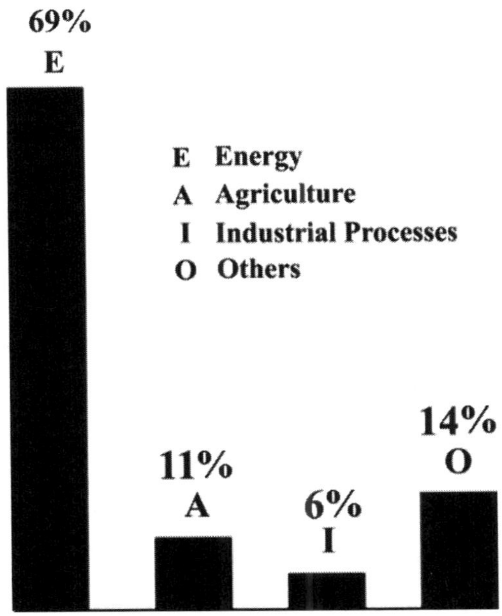

Figure 3.1 Sources of greenhouse gas emissions of the world (IEA, 2014).

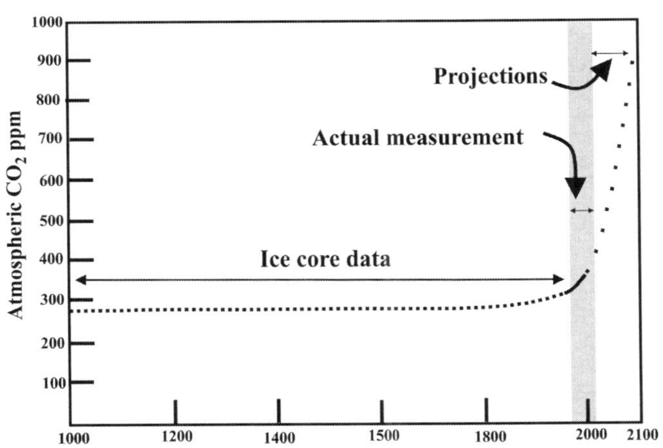

Figure 3.2 Past and present CO_2 emissions of the world, and projected future emissions (adapted from UNFCC, Chandrasekharam and Bundschuh, 2008). Pre-industrial CO_2 was determined using ice cores from the polar regions.

emissions are expected to cross the 700 ppm threshold by the end of this century, which, it is predicted, would causing severe climate related problems like harsh summers, flash floods, and cloud bursts causing severe calamity to the public and to property. If countries adopt business-as-usual policies, then the energy demand will be 50% higher than that was in 2012 and oil and coal demand will rise drastically to meet the faster growth. This will result in steep rises in oil and coal prices (the price of oil is expected to reach US$ 155/barrel by 2040) and CO_2 emissions will not be contained within a 2°C rise in temperature by 2040 (IEA, 2013).

Under the 450 Scenario countries will control use of fossil fuels as the main source of energy and the envisaged energy trajectory would limit the rise in atmospheric CO_2 to 450 ppm and, in consequence, the temperature rise to 2°C. Under this scenario the concentration of CO_2 in the atmosphere will stabilise at 450 ppm by 2040. To discourage the CO_2 emissions, carbon tax will be implemented by all the Organization for Economic Cooperation and Development (OECD) countries at US$ 140/tonne of CO_2 emitted between now and 2020. Under the BAU scenario the oil price may stabilise at US$ 100/barrel by 2040. Because of stringent control on CO_2 emissions, the coal price will fall drastically and the world will follow an energy growth rate of 0.6% by 2040 to limit the atmospheric temperature rise to 2°C (IEA, 2013). However, the New Policies Scenario being recommended for controlling CO_2 emissions takes a middle path between 450 Scenario and the business-as-usual policy. Under the New Policies Scenario recommendations, countries are expected to adopt energy efficiency methods to reduce the use of fossil fuels; to encourage the development of renewable energy sources; and to reform energy subsidies and phase out nuclear power. Both the New Policies and the 450 Scenarios recommend the removal of subsidies. Thus, by adopting the New Policies, countries will achieve a reduction in CO_2 emissions and achieve steady progress (IEA, 2013). The New Policies Scenario may be suitable for countries that have already achieved a steady state of development, but this will retard the development rate of non-OECD countries.

Whatever the policies may be, the countries should have 'self-realization' to mitigate climate change-related issues that revolve around energy and CO_2 emissions (Figures 3.3 and 3.4). But the COP are not able to come to any consensus with regard to CO_2 emissions. For example, in the Cancun meeting held in Mexico in 2010, 43 countries pledged to mitigate CO_2 emission target (to keep the CO_2 level at 450 ppm) and 48 countries pledged to initiate measures to control CO_2 emissions (Figure 3.4). However, in the 2012 UNFCC, Convention of Parties-18 have delivered an extension to the Kyoto Protocol to 2020. Under this new protocol, attended by 195 countries, 38 countries committed to control CO_2 emissions (with a share of only 13% GHG emissions) (Figure 3.3).

These meetings and conventions are not resulting in any concrete steps to control GHG emissions and the rise in temperature. Those countries that are conducting their business-as-usual are emitting huge volumes of CO_2. The CO_2 levels in the atmosphere in 2013 have already crossed 400 ppm (IEA, 2013) for the first time, and are very close to 450 Scenario of 450 ppm: a level that is expected to reach by 2020. Average global temperatures have already increased by about 0.8°C compared to the pre-industrial levels. Although carbon pricing is in force in several countries, the carbon saving prices are still at low levels in the EU Emissions Trading System (ETS). The consumption of carbon based fossil fuels, including coal, continue to rise (IEA, 2014). If this trend continues, the temperature rise is bound to touch 4.5°C in this century. It seems that the world is already experiencing extreme weather events, which would appear to be the consequences of climate changes: things like unprecedented floods, heat waves, and storm frequencies have increased in recent times. Countries are slow in implementing policies to contain the rise in global temperature to 2°C. However, it is still possible,

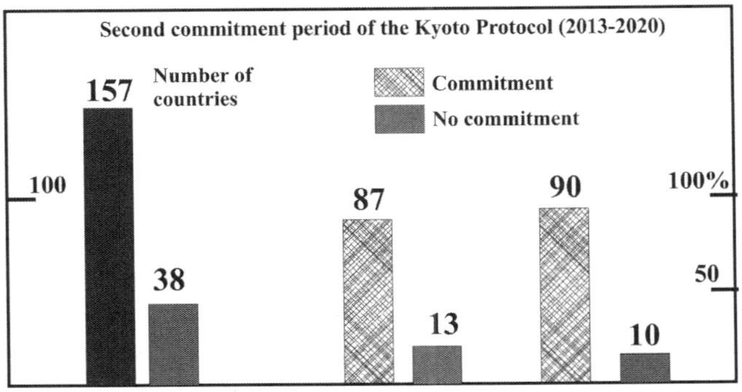

Figure 3.3 195 countries participated in the second commitment period meeting and only 38 countries committed to control GHG emissions. The share of world GHG emissions by the 157 countries in 2010 was 87% and in 2020 it is projected to be 90%. In the case of those 38 countries committed to reduce GHG emissions, the share of world GHG emissions of these 38 countries in 2010 was 13% and the share of world GHG emissions projected to 2020 will be 10% (adopted from IEA, 2013).

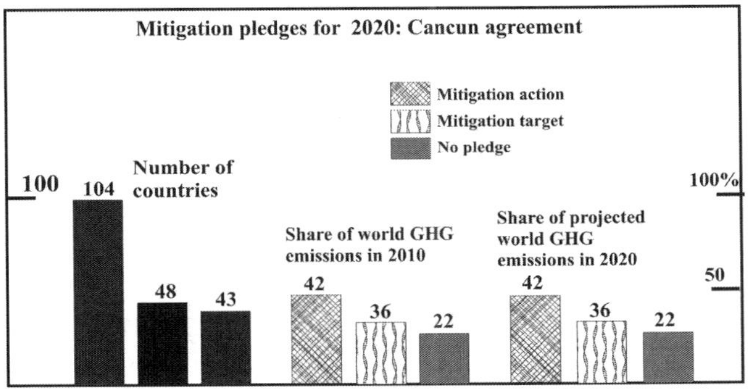

Figure 3.4 In the Cancun agreement, 195 countries attended the meeting and 104 countries did not pledge to reduce GHG emissions. 48 countries have taken mitigation to control GHG emissions while 43 countries pledged GHG emissions mitigation targets. The share of world GHG emissions by the 104 countries in 2010 was 22% and the share of projected world GHG emissions share emissions will be the same. The share of world GHG emissions in 2010 by the 48 countries that have taken mitigation action was 42% and under the projected world emissions year in 2020 this will be the same percent. In the case of 43 countries that pledged mitigation targets, the world share of GHG emissions in 2010 and in projected year of 2020 will be 22% (adopted from IEA, 2013).

technically, to limit CO_2 emissions and limit the global temperature rise to 2°C. Since energy generation is the main cause of large CO_2 emissions (Figure 3.1), switching to clean energy sources like geothermal, a source that can supply baseload electricity, will reduce the accumulation of large CO_2 volumes in the atmosphere (IEA, 2013).

From the forgone discussions it is apparent that countries may not be able to arrive at a tangible agreement to meet the target threshold of 2°C temperature rise by 2020. The World Energy Outlook (WEO, 2014) proposed pragmatic policy actions that may be taken by the countries across the world to curtail GHG emissions and maintain the

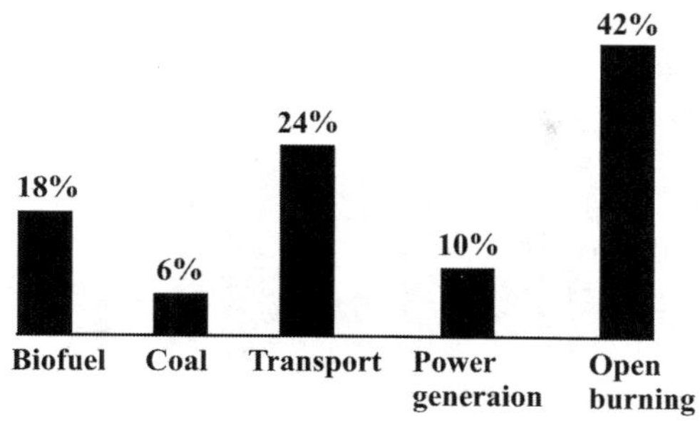

Figure 3.5 Black carbon emission by source (adopted from Streets et al., 2004, Bond et al., 2007).

CO_2 levels at 450 ppm by 2020. The WEO came out with four priority achievements. They are: 1) evolve energy efficiency measures, 2) discourage the use and construction of inefficient coal-fired power plants, 3) bring down the level of methane emissions by oil industries, and 4) partially phase out of fossil fuel subsidies. These four strategic measures, known as 4 for 2°C will, it is anticipated, account for 1) a 49% reduction resulting from energy efficiency measures, 2) a 21% reduction from discontinuation of inefficient coal-fired power plant, 3) a 18% reduction from savings from oil and gas companies (methane emissions reduction), and 4) a 12% reduction from partial phasing out of subsidies. Further, these policy measures will encourage the growth of renewables by increasing their contribution share of primary energy from 20% now to 27% by 2020 (WEO, 2014). Along with the implementation of the 4-for-2°C Scenario policies, countries should continue their negotiations to bring out a viable long-term solution to limit the GHG emissions and achieve the 2°C target by 2020. But the projections suggest that continuing with CO_2 and methane emissions at the current rates of emission will mean reaching 3.9 gigatonnes (Gt) by 2020, which will be above the levels stipulated on the presumption that this will contain the temperature rise at 2°C.

However, all the negotiations carried out, and the energy polices recommended by countries and other organisations focus only on CO_2 and GHG emissions, and there is no focus on the reduction of black carbon (BC) levels. BC is equally detrimental for global climate change and an increase in the surface temperature of the earth. BC is strongly related to the combustion of biomass for energy, and more than half the global population depends on biomass for its energy.

Black carbon (BC) emissions from anthropogenic sources (burning fossil fuels and biomass such as dung, crop residue, and forest wood) are concentrated in the tropics where the solar irradiance is highest. BC mixes with other aerosols and forms transcontinental plumes of atmospheric brown clouds that extend vertically up to 5 km. BC is the second strongest contributor to global warming (Ramanathan and Carmichael, 2008). The global annual emissions of BC are about 8 terragrams per year (Tg/yr), contributed by biofuels (20%), fossil fuels (40%), and open biomass burning (40%). Until 1950, North America and Europe were the main BC-emitting countries, but now nearly 60% of BC comes from rural areas in Asian and African countries where biofuels and biomass are the main sources of energy (Novakov et al., 2003, Streets et al., 2004, Bond et al., 2007). The BC emissions by source are shown in Figure 3.5.

3.2 EGYPT

Egypt's energy requirement is mostly met by natural gas. Egypt has 76 trillion cubic feet of remaining gas reserves and ranks seventh among the non-OPEC countries. The country still has 20 trillion cubic feet of unexplored natural gas (de la Vega, 2010). The country is planning to expand its exploration and production of natural gas to meet the world demand. From the current 600 billion cubic feet of gas exported per year, the country is planning to escalate this quantity to 800 billion by 2030. Though oil and gas are GHG emitters, gas emits less CO_2 (193 kg CO_2/MWh) relative to oil (817 kg CO_2/MWh). Energy demand in Egypt is increasing rapidly due to an increase in population, economic growth and an increase in the transport system (de la Vega, 2010). Egypt's population is expected to exceed 100 million by 2030 from the present 82 million. In a recent joint Egyptian-German committee report on renewables, de la Vega (2010) analyses energy demand for Egypt under two extreme scenarios. In the first scenario, known as the business-as-usual scenario (BAU) it is assumed that Egypt's GDP will grow at the rate of 3.1% per year between now and 2030. In the case of the second scenario, known as the high economic growth scenario (HEG), the average annual economic growth is assumed to be 4.5%. A third scenario is also analysed by the above author, known as the substitution and efficiency scenario (S&E), where renewable energy use will be greater and energy efficiency policies will be implemented to reduce consumption of fossil fuel based electricity thereby decreasing CO_2 emissions. Because of the implementation of energy efficiency policies under S&E, the energy demand decreases substantially from that of HEG (Figure 3.6). According to this analysis, S&E effects will be seen only after 2020, when there will be decline in fossil fuel use and renewables will contribute to a large extent. Apparently, under the S&E scenario, Egypt will reduce CO_2 emissions and the effect may be seen only after 2030. The energy consumption by different sectors (e.g. industry, transportation, residential, and others) do not show much variation between these three scenarios (de la Vega, 2010). Substantial improvement can be seen only under S&E, especially in the residential sector, perhaps due to the use of renewables for space cooling and for electricity (Figure 3.6).

Egypt's emission of CO_2 from oil is about 90 million tonnes (Mt), and from gas it is 88 million tonnes (Table 2.2).

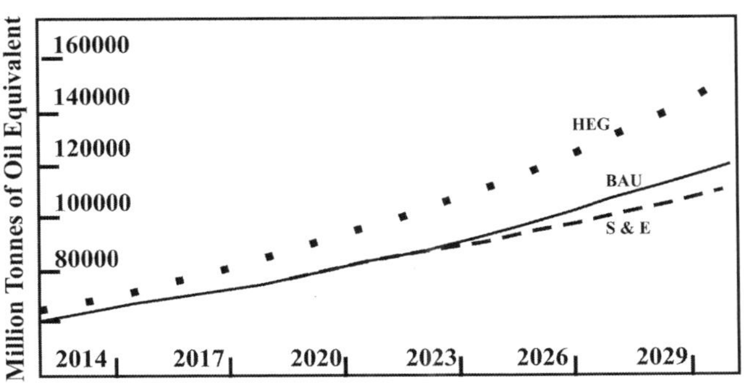

Figure 3.6 Projected energy demands in Egypt under the three scenarios (adopted from de la Vega, 2010).

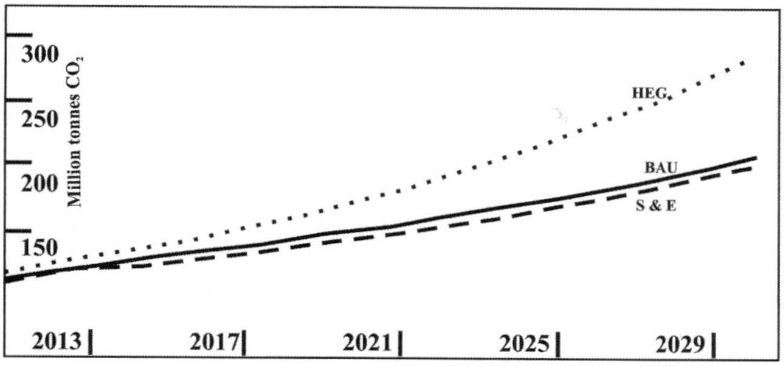

Figure 3.7 Projected CO_2 emissions for Egypt under the three scenarios (see text; there is a variation of emission values of CO_2 and the variation is within 10%. These values have no drastic influence on the inferences drawn in the text).

The average per capita emissions of CO_2 is about 2.6 tonnes (Table 2.2). Under the BAU scenario Egypt's total CO_2 emissions will reach 300 million tonnes by 2030 that is a cause for concern for the COP 18 and the policy proposal suggested by the World Energy Outlook (WEO, 2014). However, under the BAU and S&E scenarios, emissions more or less maintain the same trend, controlling the CO_2 emissions to 200 million tonnes by 2030 (Figure 3.7), although the oil consumed during 2030 is about 20,000 million tonnes less under S&E scenario, due to the influence of energy efficiency measures, if adopted by Egypt, and the large contributions from renewables if developed (Figure 3.6). This is because the CO_2 savings are mostly contributed by residential sector, while the transport and industrial sectors' contribution to CO_2 savings is marginal (Figure 3.8a and b).

It is estimated that the transportation sector will be responsible for about 100 million tonnes of CO_2 emissions in 2030 compared to the residential sector that will account for about 23 million tonnes of CO_2 emissions in 2030. This is due to link between the projected increase in population, and GDP, and the BAU and HEG scenarios. In the case of S&E, energy efficiency and alternative energy sources play significant roles in offsetting fossil fuel based growth. Substantial CO_2 savings from adopting an energy efficiency policy and using a large percentage of renewables to support the energy demand due to population growth and to sustain a healthy GDP, would have a larger impact on the global warming issue. In fact, the S&E strategy would help to build a healthy domestic and global environment.

3.3 ERITREA

Among the countries around the Red Sea, Eritrea is the least CO_2-emitting country. Eritrea has been able to decrease CO_2 emissions by 100,000 tonnes between 2010 and 2015 (also see Table 2.2), with its per capita emissions of CO_2 at 0.1 tonne due to low per capita electricity consumption (Table 2.2) compared to other countries. In Eritrea, only 20% of the population has electricity and the primary energy supply, accounting to 66%, comes from biomass (Nakicenovic and Swart, 2000). The fuelwood consumption is around 2.5 kg per day per person (IEA, 2014c). Almost all

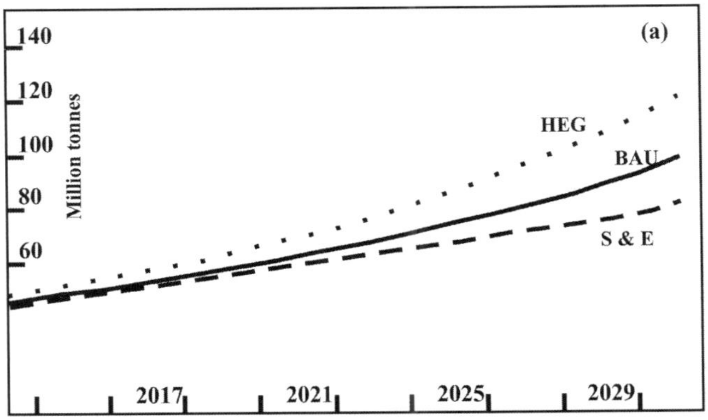

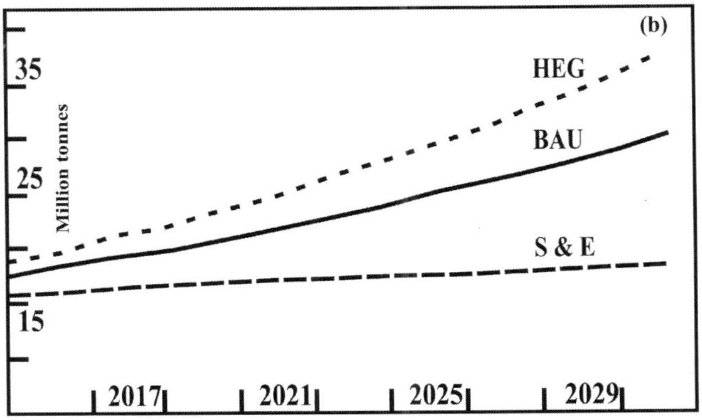

Figure 3.8 Projected CO_2 emissions in Egypt by: (a) transport and (b) residential sectors under the three scenarios (WEO, 2014).

this energy is consumed by the residential establishments and only a small percentage of this energy supports transport and industry sectors. The fuelwood consumption by Eritrea is much higher compared to the consumption by Ethiopia. Among the countries around the Red Sea, Eritrea consumes maximum amount of fuelwood, amounting to 15 million kg per day.

3.4 ETHIOPIA

Among the countries around the Red Sea, Ethiopia's population is the largest – at about 92 million – and with the highest density. However, the per capita power consumption is much lower compared to Djibouti's, whose population is only 0.9 million (Table 2.2). This could be because of a lower proportion of the population living in urban areas and, furthermore, only 23% of the population has access to electricity. Nearly 70 million people in the country still depend on traditional solid biomass to meet their energy needs (IEA, 2014c). As a result, the per capita CO_2 emissions are far lower compared to Djibouti's. However, what is not accounted for in the CO_2

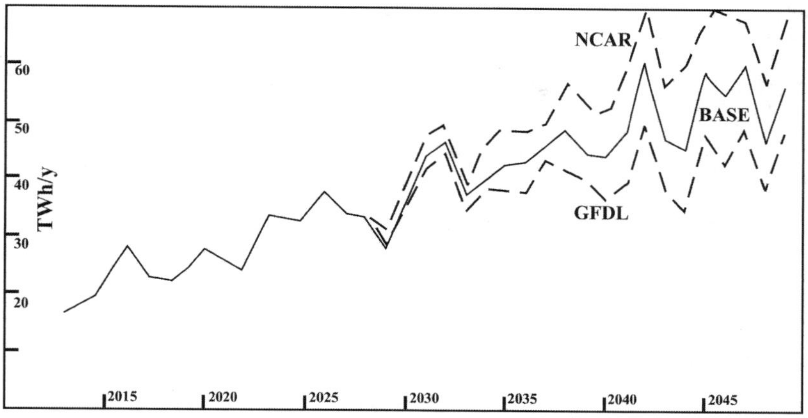

Figure 3.9 Climate change, and hydropower generation in Ethiopia: predictions. NCAR: National Centre for Atmospheric Research, GFDL: Geophysical Fluid Dynamic Laboratory (adapted from Robinson et al., 2013).

emission statistics is the amount of carbon that is emitted from biomass combustion. Since this energy usage occurs in the rural sector, it is difficult to assess the quantity of carbon emitted from this source. Perhaps an assumption can be made, considering the total CO_2 emissions and the emissions from oil sources as shown in Table 2.2. However, the emissions are considerable compared to the countries like Eritrea and Djibouti. Ethiopia depends heavily on hydropower sources for food and energy. Due to its variable topography, the country is prone to adverse impacts of climate change. Although the country experienced major droughts between 1999 and 2004, and major floods between 1999 and 2006, regional projections based on climate models suggest substantial rise in mean air temperature in the current and subsequent centuries. This is expected to cause variation in rainfall and increase the frequency of floods and droughts. These changes are going to drastically affect the country's agricultural sector (Dercon et al., 2005, Diao and Pratt, 2007, IAPAC, 2007, Robinson et al., 2013). CO_2-related climate models were developed by the National Centre for Atmospheric Research (NCAR) and Geophysical Fluid Dynamic Laboratory (GFDL) as a part of the planning process for hydropower generation development on the Ethiopian Nile (IMPEND). These models are shown in Figure 3.9.

Although Ethiopia experiences vagaries of rainfall, which would affect hydropower generation, the models show that this power generation scenario is going to be severe beyond 2030 due to variation in the air temperature caused by CO_2 emissions. As discussed in the later chapters, Ethiopia has other energy options with low CO_2 emissions to mitigate these adverse conditions. Implementing these options may, in fact, help to generate higher amounts of hydroelectric power, thus help the country to circumvent the onslaught of poverty in the next century.

3.5 DJIBOUTI

Although Djibouti has a much smaller population than Eritrea, its per capita power consumption is higher by a factor of 8 compared to Eritrea. The main sites of

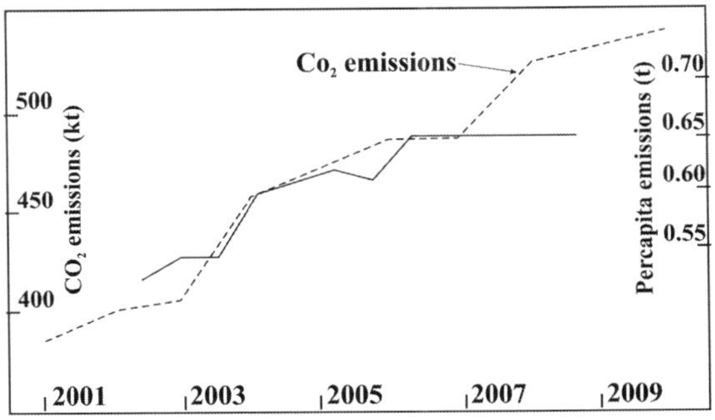

Figure 3.10 Total and per capita CO_2 emissions for Djibouti (IEA 2014c).

power consumption are confined to Djibouti District, while the rural population has no electricity. Although Djibouti has sufficiently large geothermal resources located in the regions of Lake Asal and Lake Abhe, these resources are yet to be developed and the country's energy requirement is supported by 100% imports of oil. Djibouti's has the smallest population among the countries around the Red Sea (Table 2.2). Even though the country's population is only 0.9 million, the per capita electricity consumption and per capita CO_2 emissions (Figure 3.10) are very high compared to Eritrea, whose population is greater by a factor of 8 (Table 2.2); the entire CO_2 emissions in Djibouti come from oil combustion.

Half of the country's population is concentrated in Djibouti City and only this section has access to electricity. Because of 100% imported petroleum products to fuel power plants, the cost of electric power is very high and is around US$ 0.32/kWh compared to US$ 0.1/kWh in Ethiopia. Although the installed capacity is 119 MWe, the city gets only 40 MWe of uninterrupted power supply. An additional 35 MWe is supplied to Djibouti by Ethiopia. In summer, this energy is inadequate and the demand rises to 65 MWe (Asian Development Bank, 2013).

While the per capita CO_2 emissions more or less stabilised due to low population growth (1.5% per year), the high CO_2 emissions beyond the year 2009 could be due to industrial activity like cement manufacturing. Until recently Djibouti was importing 0.12 million tonnes of cement per annum, but now several cement factories stared functioning and currently the country is importing only <0.09 million tonnes per annum (US CIA, 2000, USGS 2002). In the manufacturing of cement, 222 kg of Carbon/ton(C/t) of cement is generated (Worrell et al., 2001). This could be the reason for high total CO_2 emissions, while the per capita CO_2 emissions have been at a plateau since 2006 (Figure 3.10).

3.6 REPUBLIC OF YEMEN

The population of the Republic of Yemen is more or less the same as that of Saudi Arabia, but the density of population is much higher than in Saudi Arabia. The per capita power consumption is much higher compared to Ethiopia, which has four times the population of Yemen. This indicates that there are more urban centres in Yemen (Sana'a, Taiz, and Aden) compared to Ethiopia (Table 2.2). Consequently, the per capita CO_2 emissions are ten times higher compared to Ethiopia's. The per capita CO_2 emissions rose from 0.84 tonnes in 2000 to 1 tonne in 2011 (IEA, 2014). Electricity and heat production, and the transport sectors account for 50% of the emissions, followed by a small amount (3.1 million tonnes) contributed by the manufacturing sector (IEA, 2014). In order to reduce CO_2 emissions to mitigate climate related issues, the Ministry of Electricity and Energy (MEE, 2009) has proposed reducing oil imports and increasing the proportion of renewable energy sources used. The ministry proposed several options to meet the growing demand for electricity, and to control CO_2 emissions. These option include: a) interconnection with the national grids of neighbouring countries like Saudi Arabia and those of East Africa (e.g. Djibouti and Ethiopia), b) bilateral agreements for medium-term energy security with energy-surplus countries, c) evolvement of a long-term strategy to export carbon-free 'green' electricity, generated through renewable energy technologies, to neighbouring countries as well as countries within the European Union, and d) promote wind, geothermal, and landfill gas energy sources and concentrated solar power sources (CSP). The ministry has projected that the country will be able to generate about 35,000 MW from wind, 2,900 MW from geothermal, 6 MW from landfill gas and 19,000 MW from CSP (MEE, 2009). If these strategic plans are implemented, then the Republic of Yemen will be the first country around the Red Sea to achieve zero carbon emissions and energy security. Besides these strategic policies and plans, the ministry plans to implement energy efficiency and a demand-side management policy to further reduce CO_2 emissions by saving at least 650 GWh of electricity that is being generated at present from fossil fuels. The target year for completing the implementation of these strategic energy policies is 2025. However, due to the present political situation implementation of these plans will be delayed.

3.7 SAUDI ARABIA

Among the countries around the Red Sea, Saudi Arabia emits the most CO_2 (Table 2.2). From 297 million tonnes in 2000, the emissions jumped to 465 million tonnes in 2013. Unlike other countries around the Red Sea that use biomass to meet their energy requirements, Saudi Arabia depends entirely on its oil and gas resources.

Power plants are the major source of CO_2 emissions, followed by the industrial and transport sectors (Table 3.1). There are other sources that contribute significant amount of CO_2. These include international marine and aviation bunkers, which together contribute 15.7 million tonnes of CO_2 (IEA, 2013, 2014). The bunkers' CO_2 contribution increased from 12.4 million tonnes in 2000 to 15.7 in 2013. Within a span of two years, for example, the electricity consumption has increased by 50 TWh and the per capita electricity consumption has increased by 700 kWh (Table 3.2). With the current prevailing political situation, this amount will grow in the coming decade.

With a 6% annual increase in population growth, the demand for electricity is growing at the rate of 7.5%/year (Figure 2.13), (Chandrasekharam et al., 2014b). CO_2 emissions are a cause for concern and their effect is already felt in the ambient temperature of Saudi Arabia. The ambient temperature over the past decade over Saudi Arabia

Table 3.1 Sectoral CO_2 emissions, Saudi Arabia (in million tonnes; data source: IEA, 2013, 2014, Chandrasekharam et al., 2014a, 2015a).

Total from Fossil fuels	*Electricity and heat*	*Industries*	*Transport*	*Others*
465	206	134	120	5

Table 3.2 Oil production, electricity generation and CO_2 emissions between 2011 and 2013 by Saudi Arabia (adapted from IEA, 2013, 2014, Chandrasekharam et al., 2014a, b, 2015a, b, c).

	Oil (Million t) Production	*Gas (bcm) Production*	*Electricity (TWh) from oil*	*Electricity (TWh) from gas*	*Population Million*	*TPES Million t Oil e*	*Electricity Conp. TWh*	*CO_2 emission Mt of CO_2*	*Elec. Consump. kWh/capita*	*CO_2 emission t CO_2/Capita*
2011	544.00	95.00	142.00	112.00	28.10	187.00	227.00	458.00	8068.00	16.28
2013	540.00	84.00	150.00	121.00	28.80	200.00	271.00	465.00	8763.00	17.22

TPES: Total Primary Energy Supply, conp: and consump: consumption.

has increased by 0.7°C (Almazroui et al., 2012). Such temperature variations are intimately related to the concentration of CO_2 in the atmosphere and several countries are adopting suitable policies and mitigation strategies to reduce emissions to control the increase in the ambient temperature. As reported above, Saudi Arabia and Ethiopia are experiencing unprecedented rain, floods, and droughts (Dercon et al., 2005, Diao and Pratt, 2007, IAPAC, 2007, Robinson et al., 2013). These countries are caught in a vicious circle: with change in ambient temperature, additional oil and gas sources are used for space cooling and space heating, which will further enhance the CO_2 emissions. Currently, Saudi Arabia utilises 80% of its the electricity for space cooling. The country consumed 240 TWh of electricity in 2010, generated from oil and gas, and the projections are that electricity generation capacity will reach 736 TWh by 2020 (IEA, 2012, WB 2009, Chandrasekharam et al., 2014c) thereby consume 500,000 barrels of oil. In summer this number will reach 900,000 barrels (about 126 million kg of oil) due to demand in space cooling applications in residential and commercial establishments. This additional oil usage can generate about 1.5 million MWh of electricity thereby emitting a million tonnes of CO_2 (oil based power plants emit 817 kg of CO_2/MWh), (Chandrasekharam and Bundschuh, 2008). These additional CO_2 emissions are not accounted for in the emissions statistics published by leading organisations. Saudi Arabia's CO_2 emissions from fuel combustion have increased from 252 million tonnes (Mt) in 2000 to 465 Mt (Table 3.2) in 2014, with oil contributing 291 Mt and gas contributing 152 Mt (Table 2.2) (IEA, 2012, Chandrasekharam et al., 2014b, c) from the electricity-generation sector. The current per capita emissions of CO_2 have increased to 17 Mt from 12 Mt in 2000 (Table 2.2).

In addition to electricity generation, Saudi Arabia consumes considerable fossil fuels for generating fresh water through desalination. This is essential, as the country receives scanty rainfall and has no defined drainage systems and groundwater resources. Saudi Arabia consumes 134 million kWh of electricity to generate 275 l/day through desalination (Chandrasekharam et al., 2014a, b, 2015c). About

13 million tonnes (Mt) of CO_2 is emitted from desalination plants, if oil is the fuel, and 3 Mt of CO_2 if gas is the fuel source (Chandrasekharam et al., 2015a). At present, 33 desalination plants are in operation in Saudi Arabia with a government subsidy of US$ 0.03/m^3 of water generated, which is much lower than in other countries of the world where the average cost is US$ 6/m^3 of water generated (Taleb and Sharples, 2011, Chandrasekharam et al., 2015c). In the next decade fresh water demand will grow by 20%, compelling the administration to increase the capacity of desalination plants, or install more desalination plants, thereby also increasing the CO_2 emissions (Chandrasekharam et al., 2015c). If agricultural activity is included, then the demand for fresh water escalates at an exponential rate. In order to reduce water demand in the agricultural sector, the government of Saudi Arabia has banned the cultivation of wheat and barley from 2016. Wheat and barley are the staple food of Saudis and are grown in the western part of the country along the Red Sea where irrigation is possible due to the presence of small check dams and small quantities of groundwater, as well as fresh water available from desalination.

Wheat is an important food item in Saudi Arabia, consumed in the form of bread (pita). Average per capita bread consumption is about 241 g per day. In the year 2013–14 the total wheat consumption in Saudi Arabia was 3.25 million tonnes. Due to the new government policy on growing food crops, wheat imports are expected to reach 3.03 million tonnes in 2013–14, compared to 1.92 million tonnes in 2012–13 (USDA, 2013). Thus CO_2 related climate factors are forcing the country to sacrifice its policy on food security, which is detrimental to the country. If policies to adopt CO_2 mitigation strategies are not implemented, Saudi Arabia may not be able to economically and politically sustain the process of generating fresh water through desalination using fossil fuels (Chandrasekharam et al., 2015c) and if this trend continues, then Saudi Arabia will become an oil importer in the next two decades (Ahmad and Ramana, 2014), leading to destabilisation of the global economy.

Chapter 4

Geothermal provinces

> *"Geothermal is 100 per cent indigenous, environmentally-friendly and a technology that has been underutilised for too long".*
>
> *Achim Steiner, UN Under-Secretary General and UNEP Executive Director, at the UN Climate Convention Conference in Poznan, Poland, December, 2008*

All the geothermal provinces over the landmasses around the Red Sea evolved due to the dynamic tectonic, plutonic and volcanic processes that have been operating since the Precambrian. The intensity of these processes was at its peak between 31 to 25 Ma (Bosworth et al., 2005) and these activities are still continuing at present (Moufti et al., 2013, Duncan and Al-Amri, 2013, Hamlyn et al., 2014, Koulakov et al., 2014, 2015). The Arabian-Nubian Shield (ANS) was formed due to the accretion of island arcs and an oceanic plate which were generated due to plume activity (Gass, 1981). The contacts of these separated crustal blocks are represented by suture zones enclosing ophiolite assemblage of rocks. These suture zones and the ophiolites are part of the regional mega sutures representing the convergence of the East-West Gondwana during the Pan-African tectonic regime (900–550 Ma) (Gass, 1981, Bakor et al., 1976, Rogers et al., 1995, Stern, 1994). This Neoproterozoic Arabian-Nubian Shield was later separated into the Arabian and Nubian shields by later tectono-magmatic events that gave rise to the formation of the Red Sea at around 31–25 Ma (Cochran and Martinez, 1988). The initiation of the Red Sea rift and subsequent formation of the Red Sea was triggered by the onset of volcanism over Eritrea and Afar sea floor spreading and giving rise to the East African Rift system. Basaltic and rhyolitic magmatism dominated Djibouti and Republic of Yemen. In Egypt, volcanism is represented by the Mesozoic and Tertiary basalt flows and dikes (Siedner, 1973, Meneisy and Kreuzer, 1974). The western coast of Saudi Arabia also experienced volcanic activity due to the opening of the Red Sea, represented by dike swarms, which intruded parallel to the coast with an outpouring of lava that gave rise to volcanic centres that are known as *harrats*. Volcanic activity in these harrats is still continuing (Koulakov et al., 2014, 2015). The Al Madinah basalt flows in Saudi Arabia, recorded the youngest K/Ar and Ar/Ar ages of 1.7 Ma (Moufti et al., 2013, Duncan and Al-Amri, 2013) and volcanic activity in this region continued till 1256 AD (Bosworth et al., 2005, D'Almeida, 2010). Volcanism is still active in all the countries around the Red Sea (Brown, 1970, Bayer, et al., 1989, Moufti et al., 2013, Chandrasekharam et al., 2014c). The initial rifting of the Red Sea followed the Arabian-Nubian tectonic fabric initiated by plume related volcanism towards the southern part of the Red Sea and migrated north giving rise to the Red Sea. This activity initiated between 31 and 25 Ma and was followed by sin-rift intrusive activity represented by dike swarms (Bosworth et al., 2005). This Red Sea tectonic activity resulted in an outpouring of basaltic lavas over the western Arabian Shield, giving rise to large volcanic fields or centres, which are known as "harrats". A thinning of continental crust and high heat flow values along the coast generated large volumes of granitic melts that intruded the Precambrian basement as well as the volcanic flows along the Arabian Shield. These post tectonic granitic intrusives are enriched in uranium, thorium and potassium. The entire tectono-magmatic activities around the Red Sea gave rise to several geothermal provinces over all the continents surrounding the Red Sea, represented by thermal springs with very high issuing temperatures and fumaroles in several locations in Eretria, Djibouti, Ethiopia, Republic of Yemen and Saudi Arabia (Figure 4.1).

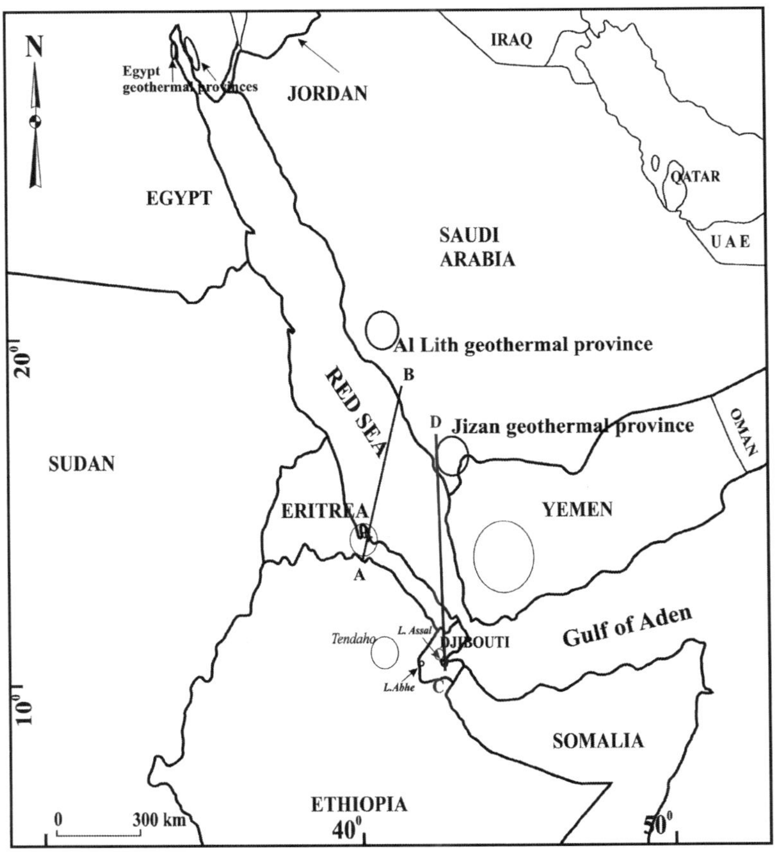

Figure 4.1 Geothermal provinces around the Red Sea. Circles indicate geothermal provinces. Subsurface lithological sections are drawn across A-B and C-D and are discussed in the subsequent sections (adapted from Chandrasekharam et al., 2014c).

The evolution of the thermal provinces from the six countries around the Red Sea is discussed in the following sections in terms of geological and tectonic features, and geothermal characteristics of the fluids.

4.1 EGYPT

4.1.1 Geological and tectonic setting

The Red Sea rift in the north bifurcates into two, the Gulf of Suez and the Gulf of Aquaba. The Gulf of Suez is the failed arm of the Red Sea rift system during the Oligocene to Early Miocene (Colletta et al., 1988, Patton et al., 1994, Zaher et al., 2011). Part of the Nubian Shield represented by the Precambrian basement is present over a large part extending along the Red Sea and southern part of the Sinai peninsula (Figure 4.2).

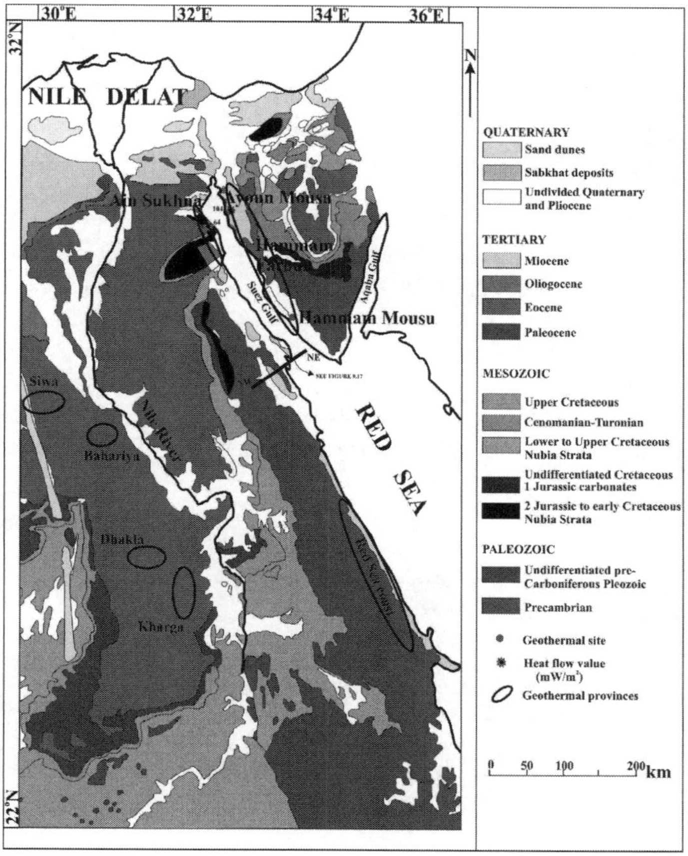

Figure 4.2 Geological and structural map of Egypt showing the geothermal provinces (adapted from Chandrasekharam et al., 2015c).

This Neoproterozoic Arabian-Nubian Shield was formed due to an accretion process under an island arc tectonic environment between 800 and 600 Ma (Stern, 1985, 1994, Stern and Johnson, 2010). The basement is characterised by greenschist facies metamorphosed volcano-sedimentary rocks associated with high grade metamorphic rocks that are common in all such tectonic environments across all the continent. A large part of the Nile flood plain is covered by the Tertiary sediments that lie over the Mesozoic and Paleozoic formations (Figure 4.2). The Precambrian rocks are also exposed in south Sinai and the depth to the basement increases towards the north. The entire Paleozoic, Mesozoic, Tertiary and Quaternary sequence that is present in the Saudi Arabian continent is also exposed in Sinai and the western margin of the Red Sea in Egypt (Saleh et al., 2006. Figure 4.2). Extensional rifting within the Gulf of Suez influenced by the active Red Sea rifting resulted in several horst and graben structures along the eastern and western margins of the Gulf of Suez. Early regional tectonic regime in Egypt is represented by two mega lineaments that cut across the African continent and truncate towards the Red Sea in Egypt. The Trans-African Lineament (TAL) extends from Nigeria in western Africa to the Nile delta in Egypt, and the Central African lineament (CAL) that extends from Cameroon in western Africa to the Red Sea hills in Sudan (Nagy et al., 1976, Schandelmeier and Pudlo, 1990, El Ahmady Ibrahim et al., 2015). These two mega trans-continental structures cut across the Precambrian Shield in Egypt. These two major shear zones are the loci for alkaline magmatism represented by alkali granites that are weathered and classified as silicified and kaolinised granites (Eby, 1992) that are enriched in rare earth elements (REE) and radioactive elements like uranium and thorium. These granites are associated with rhyolitic lava flows, rhyolitic tuff, spherulitic rhyolites, trachyte and rhyolite dike swarms (Elsayed et al., 2014). These alkaline granites sometimes show intrusive contact with the volcanics and the volcanics at places contain these granites as enclaves. The entire alkaline suite contains a high content of uranium and thorium that varies from 13 to 300 ppm and 12 to 1100 ppm respectively (Elsayed et al., 2014). In addition to this alkaline suite of rocks, the Precambrian basement complex, in Abu Ziran in the eastern desert of Egypt, is intruded by two sets of granites: calc-alkaline granites belonging to 800 to 610 Ma (older) and alkaline granites belonging to 600 to 530 Ma (younger) (Elsayed et al., 2014). The uranium and thorium content in these granites varies from 1.2–2.9 and 5 to 24 ppm respectively (Table 4.1).

Table 4.1 U, Th (ppm) and K% concentration in certain alkaline plutonic and volcanic rocks and their heat generation (Source: 1: Firtz et al., 2014, 2: El Sayed et al., 2014).

	K_2O%	*U ppm*	*Th ppm*	*K (%)*	*$\mu W/m^3$*	*mW/m^2*	
Abu Ziran granite, Red Sea coast ED 106	3	2	24	2.4903	2.4	64.1	1
Abu Ziran granite, Red Sea coast ED 114	3	3	17	2.4903	2.2	61.8	1
Abu Ziran granite, Red Sea coast ED 125	3	2.4	17	2.4903	2.0	60.3	1
Rhyolite tuff	2.1	16	30	1.74321	6.4	103.5	2
Spherulitic rhyolite	3.3	28	73	2.73933	12.5	165.0	2
Alkaline rhyolite dike	2.8	39	66	2.32428	14.8	188.0	2
Peralkaline granite	6	300	1100	4.9806	153.6	1576.1	2
Volcanoclast	1.6	52	106	1.32816	20.8	248.2	2

Table 4.2 Physical and chemical characteristics of representative cold and thermal waters and Red Sea water, Egypt. (Location of the thermal springs is shown in Figure 4.2, adapted from Chandrasekharam et al., 2015c).

Samp. No	*Temp*	*Area*	*pH*	*Na^+*	*K^+*	*Ca^{++}*	*Mg^{++}*	*Cl^-*	*HCO_3^-*	*SO_4^{--}*	*$d^{18}O$*	*dD*	*B*	*F*	*TU*
Hammam Faroun 1	70	Egypt	6.5	4750	130	1039	489	9654	132	1450	−5.59	−41.3	0.2	2.5	0.5
Hammam Faroun 2	51	Egypt	7.1	4280	117	1039	422	8713	103	1450	−5.78	−41.5			
Ayoun Moussa	37	Egypt	6.6	1794	60	594	267	3768	287	1250	−6.14	−42.4	0.9	6.6	
Ain Sukhna	32	Egypt	7.6	1970	75.0	408	250	3730	207	1300	−6.7	−45.7	0	2	0
Nile R	28	Egypt	8.3	26	5.0	39	15	18	317	24	N/A				
4G Spring	28	Egypt	8.0	520	5.0	270	110	1113	300	633					
Red Sea				92850	1870	5150	764	155500	143	840					

Egypt experienced volcanic activity between Jurassic-Early Cretaceous (Mesozoic volcanics, 155–125 Ma) and Late Cretaceous-Early Tertiary (90–60 Ma). They are exposed in the eastern and western deserts and in the Sinai (Figure 4.2). The Cretaceous volcanic activity continued during Cenozoic and Palaeocene represented by pulses of volcanic activity. This was followed by wide spread Tertiary basaltic volcanism that continued in Eocene. This episode was followed by volcanic activity related to the opening of the Red Sea (Perrin et al., 2009).

4.1.2 Geothermal manifestation

Geothermal provinces in Egypt are distributed along the Red Sea coast, east of the Nile, eastern and western margins of the Gulf of Suez (Hammam Faroun, Ayoun Moussa, Ain Sukhna), around Kharga, Dhakla, Bahariya and Siwa oases (Figure 4.2). Those thermal springs on the either side of the Suez Gulf are high temperature springs (Table 4.2) while those in other localities are warm springs with temperatures varying from 35 to 42°C (Swanberg et al., 1983).

The geothermal gradient and heat flow values in these sites are above the global average values. The temperatures recorded from certain oil wells located around Hammam Faroun, Ain Sukhna and Ayoun Moussa (Figure 4.2) at 2 km depth vary from 120 to 260°C and the heat flow values recorded are >95 mW/m^2 (Morgan et al., 1976, Zaher et al., 2011, 2012). Gravity, magnetic investigations together with numerical simulations indicate deep circulation of meteoric water as the source for the thermal springs in Hammam Faroun. Deep seated faults associated with horst and graben structures, developed due to rifting, are the main channels for fluid circulation. A similar situation is expected in other geothermal sites (Zaher et al., 2011, Abdalla, and Scheytt, 2012). The majority of the thermal springs in all the geothermal provinces circulate through granites that contain a high concentration of radioactive elements. The chemical constituents in the thermal and groundwater samples are shown in Table 4.2.

The thermal waters are enriched in Cl^- ions and fall in the Na-Cl field in Langelier and Ludwig's diagram (Langelier and Ludwig, 1942) shown in Figure 4.3. However,

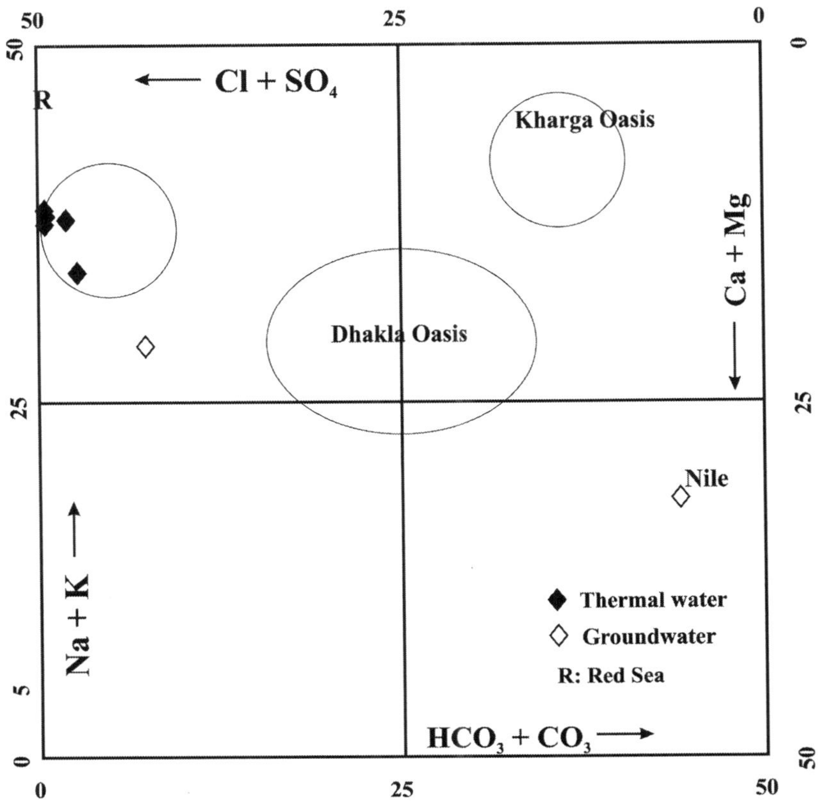

Figure 4.3 Langelier and Ludwig's (1942) major ion characteristics of Egypt's thermal and cold waters. Red Sea sample is also plotted for reference.

the samples plot away from the Red Sea, suggesting a source other than sea responsible for high chloride content. Out of the two samples of groundwater, one is a typical Ca HCO_3 type while the second sample is saline and plots near the thermal springs in the Na-Cl field. Since the thermal springs are located near the Gulf of Suez, the probability of sea water mixing with the thermal springs is high. Those thermal springs located on the western side of the Nile, in the two oasis are of different composition, indicating a mix between thermal and groundwater. The Na-HCO_3 type water recorded in Kharga Oasis appears to be emerging to the surface after circulation in rocks of granitic composition.

The position of the thermal waters around the Gulf of Suez in Figure 4.4 indicates their deep circulation and minimum mixing with the near surface cold waters during its ascent, as suggested by Giggenbach (1988), while those thermal waters from the eastern and western desert provinces indicate mixing and or thermally heated groundwaters (Figure 4.4). The fluoride content in the thermal waters from the Gulf of Suez (2 to 4 ppm, according to Swanberg et al., 1983) is typical of waters circulating in granitic reservoirs (Chandrasekharam et al., 2015b, Lashin et al., 2014). The thermal springs

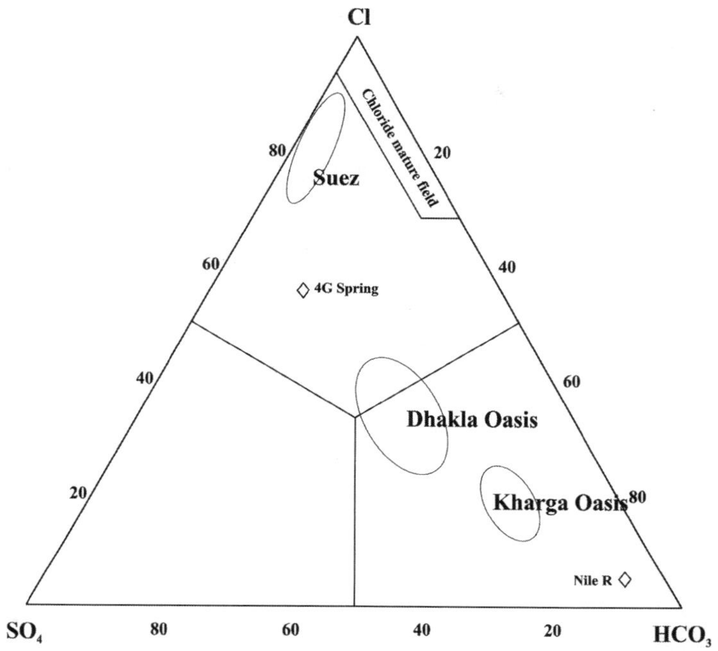

Figure 4.4 Cl-SO_4-HCO_3 diagram of Giggenbach (1988) characterising the thermal waters from Egypt. Suez thermal waters, cold and Nile river water characteristics are given in Table 4.2.

along coast of the Gulf of Suez, located within a high geothermal gradient and heat flow value region, indicate higher reservoir temperatures relative to those occurring in the oases' regions (Figure 4.5).

4.2 ERITREA

4.2.1 Geology and tectonic setting

Eritrea, located within the Danakil Depression is one of the most active regions of the Red Sea coast (Figure 4.6). This landmass is part of the Nubian Shield formed due to the activation of the Red Sea rift and Gulf of Aden spreading rift (Bosworth et al., 2005, Lowenstern et al., 1999).

Two active volcanoes are located within this depression, Musa Ali and Alid. A part of the Erta Ale volcano also falls within Eritrea (Figure 4.6). The Danakil horst separates the Red Sea and the Eritrean plateau. The Precambrian basement crystalline complex, a part of the Arabian-Nubian Shield, occupies a large landmass along the Red Sea. Subsequent to the Red Sea rift, Mesozoic-Tertiary sediments (marine sandstone, siltstone, gypsum beds and fossiliferous limestone of Pleistocene age) covered the region followed by basalt flows of Tertiary age. The continued and active rifting and subsidence in Danakil resulted in an intensity of Tertiary volcanic activity that covered the entire coastal region. The volcanic activity continues to be active today

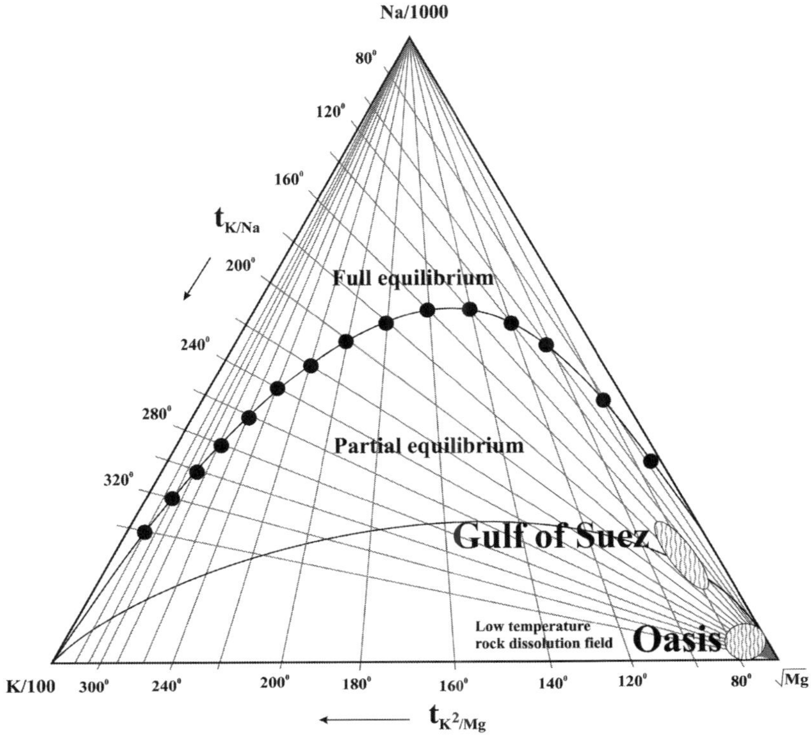

Figure 4.5 Giggenbach K-Na-Mg diagram (Giggenbach, 1988) indicating high reservoir temperatures for the Gulf of Suez thermal springs compared to other thermal springs in the western Nile desert oases regions.

(Bosworth et al., 2005, Lowenstern et al., 1999, Chandrasekharam et al., 2015c). The Alid volcanic centre in the Danakil Depression is a structural dome formed due to the intrusion of rhyolitic magma after the main Pleistocene sedimentary formation. The rhyolitic magma intrusion and subsequent rifting induced several weak linear zones that resulted in regional faults and tensional cracks along a NNW-SSE direction (Figure 4.6). These faults and tensional cracks channelised saline fluids (mainly seawater) around the thermal areole that gave rise to thermal springs and fumaroles around the volcanic centres (Duffeld et al., 1997, Chandrasekharam et al., 2015c).

4.2.2 Geothermal manifestation

Geothermal manifestations in Eritrea are represented by hot springs, fumaroles and steaming ground. The most prominent sites are located around Alid and Nabro-Dubbi volcanic centres and around the Zula Gulf (Figure 4.6). The meteoric water and the Red Sea water flowing through the fractures systems around these volcanic centres are the main feeders to the circulating geothermal fluids (Yohannes, 2007, Lowenstern et al., 1999, Duffeld et al., 1997, Chandrasekharam et al., 2015c).

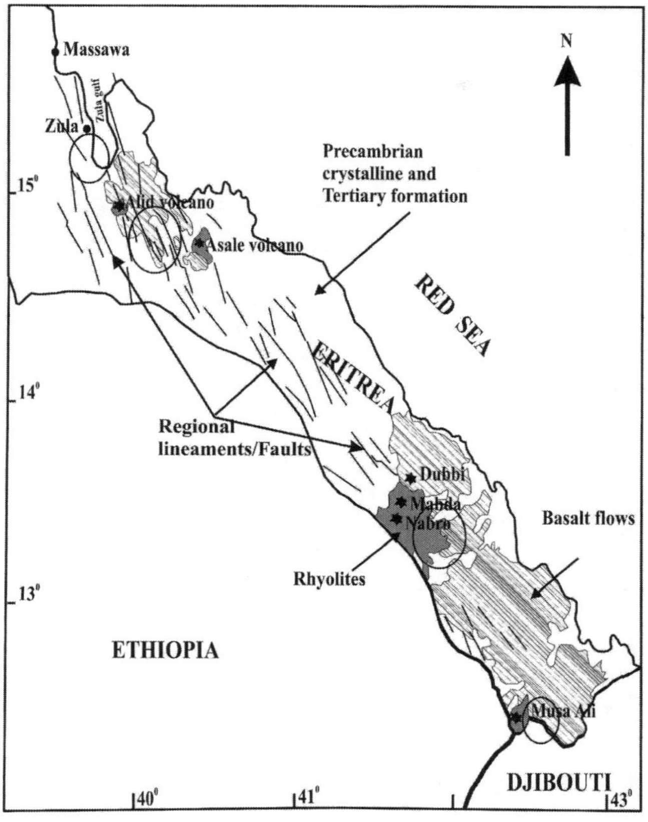

Figure 4.6 Geological and tectonic map of Eritrea. Geothermal manifestations are shown as circles (adapted from Chandrasekharam et al., 2015c).

Representative chemical analyses of the thermal and cold waters and gas condensates from Alid geothermal site are given in Table 4.3 and the data are plotted on the Langelier and Ludwig diagram to understand the behaviour (Figure 4.7). The Red Sea's signature and the influence of volcanic gases is very visible in the thermal waters. High CO_2 from volcanic emanations is reflected in the gas condensates, the thermal waters plot in the NaCl field with dilution from non-volcanic-meteoric waters. Low annual rainfall (500–700 mm, according to Duffield et al., 1997) enhances a large inflow of sea water through local and regional fault systems into the reservoirs and also influences the mixing process during the ascent of the thermal fluids (Chandrasekharam et al., 2015c).

Th thermal springs fall into three fields in Figure 4.8 indicating their high temperature signatures. Those samples falling within the Giggenbach (1988) field demonstrate the presence of high temperature reservoir in these provinces. Since the geothermal systems are associated with an active volcanic field, the presence of high temperature geothermal systems is not uncommon. Those samples plotting at the HCO_3 apex are

Table 4.3 Representative chemical analyses of geothermal and cold waters and condensates from Alid geothermal site (adapted from Duffield et al., 1997, Lowenstern et al., 1999. For sample location see Lowenstern et al. 1999).

	Samp. No	*Temp*	*pH*	*Na^+*	*K^+*	*Ca^{++}*	*Mg^{++}*	*Cl^-*	*HCO_3^-*	*SO_4^{--}*	*F*	*$d^{18}O$‰*	*dD‰*	*CO_2mol%*	*H_2Smol%*	*H_2mol%*
1	ELW96-2	58	7.2	2320.00	120.0	940.00	88.80	5134.00	30	216.00	2.87	−16	0			
2	ELW96-7	54	7.5	233.00	20.0	396.00	27.20	20.90	100	1475.00	0.50	3.2	33			
3	ELW96-8	57	7.6	213.00	17.0	251.00	21.70	12.40	66	1068.00	0.43	3.6	35			
4	ELW96-9	66	7.4	11.40	12.0	157.00	37.40	0.84	171	949.00	1.18	4.4	24			
5	EU396-2*	95	6.5	0.06	0.1	0.05	0.02	0.07	189	1.71	0.16	−4	10	98	0.23	1.01
6	ELG96-6*	94	6.7	0.23	0.1	0.50	0.03	0.07	153	0.55	nd	−2.8	−1	99	0.14	0.61
7	ErF96-1	30	7.8	508.00	9.4	134.00	122.00	747.00	302	608.00	0.77	−2.8	−9			
8	ErF96-2	34	7.6	71.10	20.0	42.60	7.60	42.00	245	25.50	0.16	−2.1	−2			
9	ErF96-3	35	7.9	713.00	12.7	59.90	31.40	841.00	548	218.00	1.80	−1.9	−1			
10	ErF96-4	35	7.4	47.60	10.5	146.00	57.90	73.00	102	496.00	2.77	−1.4	4			

*Gas condensate; nd: not detrmined

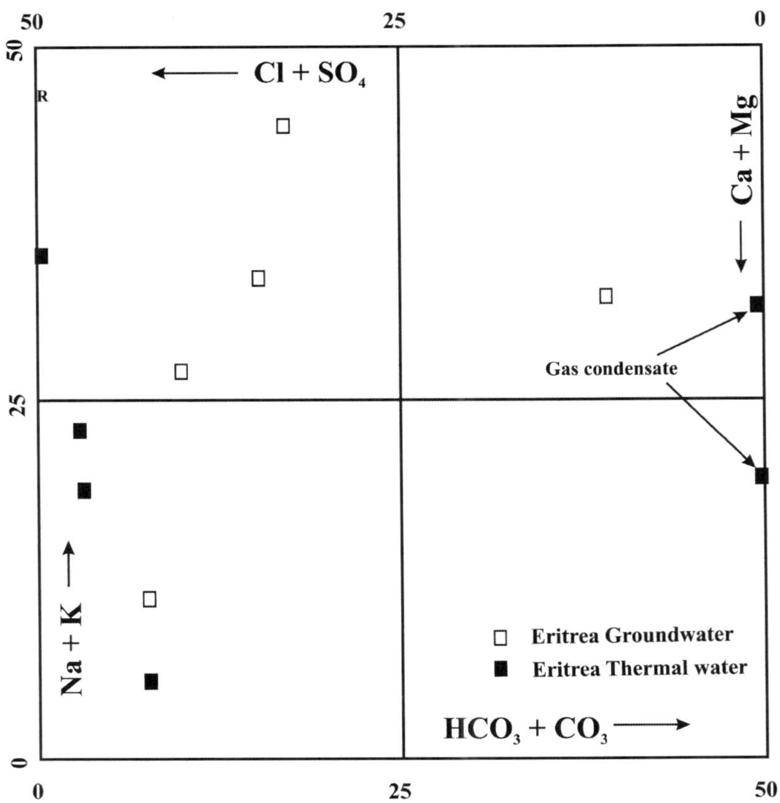

Figure 4.7 Langelier and Ludwig (1942) diagram showing the chemical characteristics of thermal, cold waters and condensates from Eritrea. Red Sea sample is also plotted for reference (adapted from Chandrasekharam et al., 2015c).

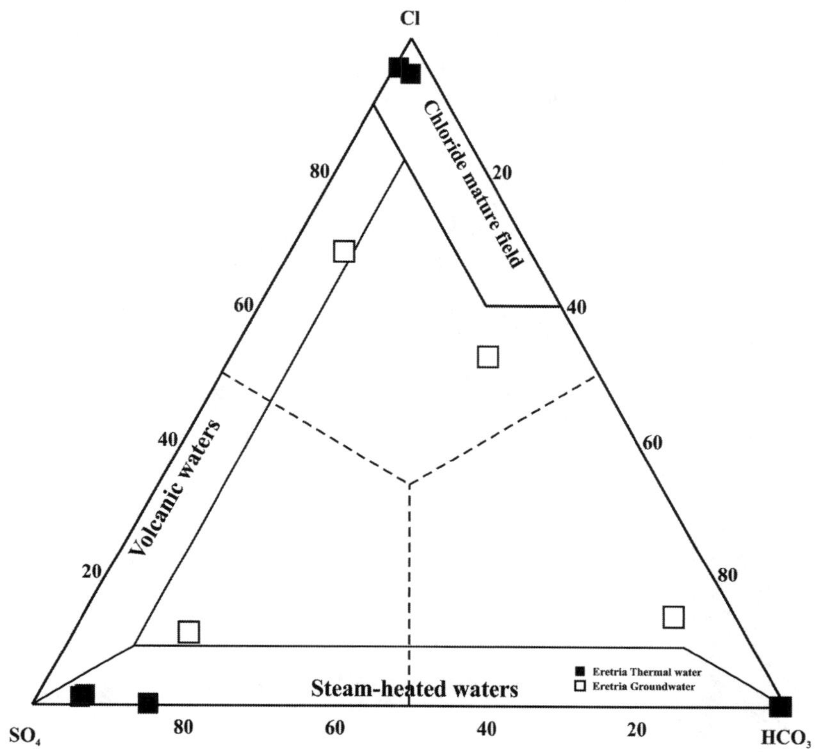

Figure 4.8 Cl-SO_4-HCO_3 diagram (Giggenbach, 1988) showing the chemical diversity of the thermal and cold waters (adapted from Chandrasekharam et al., 2015c).

steam condensates while SO_2 dissolved thermal waters fall towards the SO_4 apex in Figure 4.8 (Chandrasekharam et al., 2015c). The cold water positions in Figure 4.8 represent mixing with thermal waters and a few appear to be steam heated subsurface waters. Steam condensates and steam heater groundwater do not offer any information related to the reservoir and fall towards the Mg corner in Figure 4.9, while the thermal waters fall within the partial equilibrium in Figure 4.9. Thus, the thermal waters' signature in Figure 4.9 also confirms the presence of a high temperature geothermal system in this region.

It has been reported that the hydrothermal systems operating in this region will have temperatures varying from 250 to 300°C (Duffield et al., 1997) but the temperatures shown in Figure 4.9 are less than 250°C. However, as discussed in the later sections of this chapter, the thermal waters show positive oxygen isotope shift and the steam condensates fall on the left side of the meteoric line in the oxygen-hydrogen isotope diagram indicating an exchange of oxygen isotopes between the rocks and the thermal fluids. Oxygen isotope exchange between rock and water is facilitated at temperatures at or above 220°C (Nuti, 1991).

The chemical and isotope data indicate the geothermal system in and around Alid volcanic centre is located just above the rhyolitic dome located at a depth of about

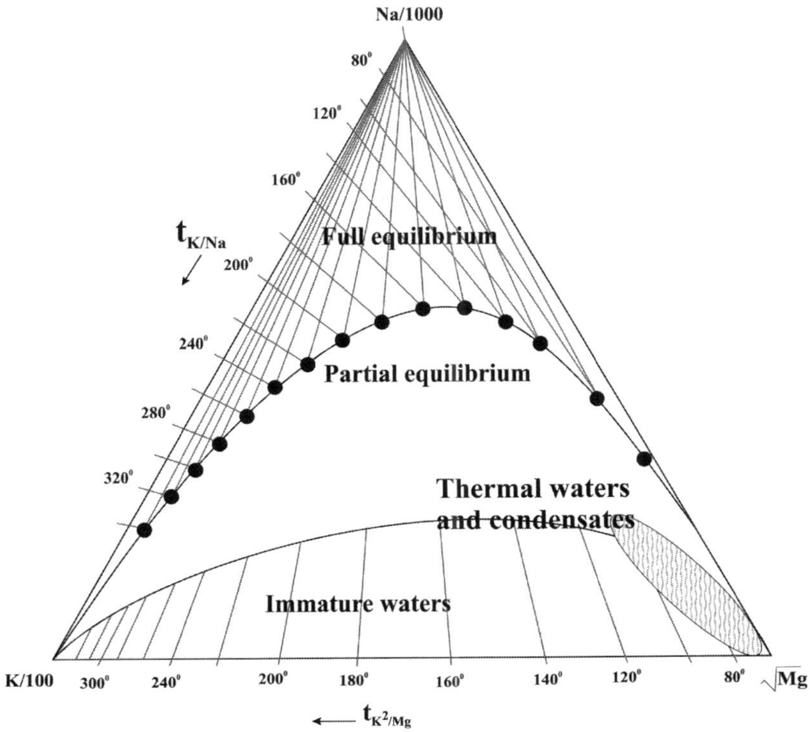

Figure 4.9 Na-K-Mg diagram (Giggenbach, 1988) indicating the temperature of the geothermal systems, Eritrea (adapted from Chandrasekharam et al., 2015c).

500 m and that the rhyolitic dome is located at about 3 km. The geothermal reservoir appears to be vapour dominated located above the rhyolitic dome intruded into the volcanic cone. The subsurface beneath Alid volcanic centre is highly fractured and hence developed a high degree of permeability giving rise to a well-developed geothermal systems (Lowenstern et al., 1999).

4.3 ETHIOPIA

4.3.1 Geology and tectonic setting

The East African Rift system is one of the most active tectonic and magmatic rift systems in the world. This region exhibits an ongoing continental breakup process associated with large scale volcanism. The Ethiopian rift, which forms a part of the East African Rift system, is the result of an extension of the triple junction i.e. the Red Sea, Aden and Ethiopian rift systems. The initial Ethiopian rift started at about 18 to 15 Ma, with the development of a NE directional seafloor spreading (extension of the Gulf of Aden seafloor spreading) (Wolfenden et al., 2004, Bosworth et al., 2005). This tectonomagmatic activity resulted in geothermal manifestations at several sites along the rift

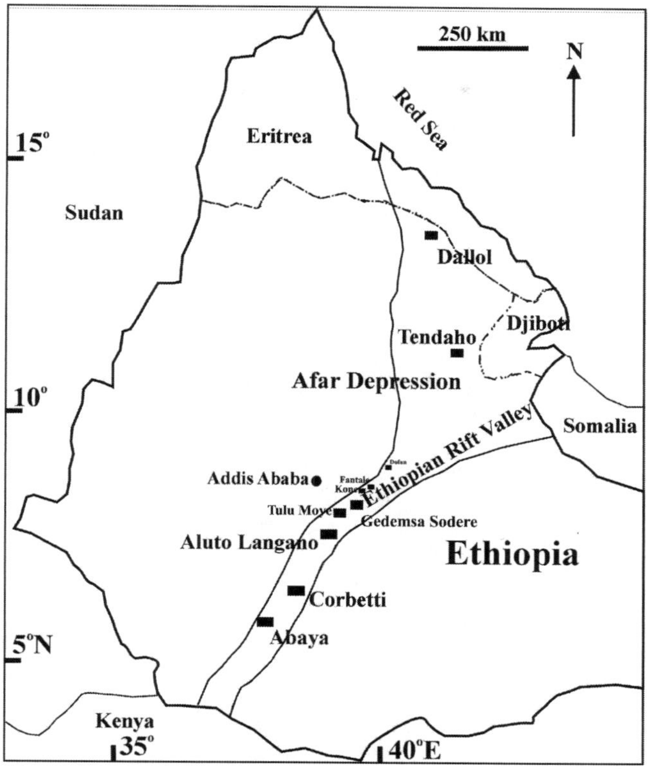

Figure 4.10 Geothermal provinces of Ethiopia (adapted from Chandrasekharam and Chandrasekhar, 2010).

floor. The most significant geothermal sites within the rift floor are Dallol, Tendaho (Dubti), Aluto Langano, Corbetti and Abaya (Figure 4.10). The main Ethiopian rift, which commenced between 18 to 15 Ma (Woldegabriel et al., 1999, Wolfenden et al., 2004), lies over the Ethiopian plateau that developed over a mantle plume which caused extensive volcanism over all the land masses around the Red Sea and the opening of the Red Sea and breakup of the Arabian-Nubian Shield. Radiometric data on the rocks from the Red Sea and Gulf of Aden indicate extensive basaltic and felsic volcanism over an area of 1000 m diameter which coincided with the opening of the Red Sea (Wolfenden et al., 2004, Bosworth et al., 2005). A detailed account of the evolution of the continents around the Red Sea is given at the end of this chapter. The most explored geothermal sites in the main Ethiopian rift valley are Dallol, Tendaho, Aluto Langano and Corbetti. Only the Tendaho geothermal site details (Figure 4.10), close to Lake Abhe, are included in this book. The chemical characteristics of the thermal and cold waters of Tendaho are shown in Table 4.4.

The Tendaho geothermal province is located N-NW of the Lake Abhe geothermal province in Djibouti and on the NW side of the Damale volcano. The Tendaho geothermal province extends over an area of 2500 km^2. The Awash River that originates from the catchment area on the southern part of the Main rift, flows along the Tendaho

Table 4.4 Chemical and isotopic signatures of thermal and cold waters from Tendaho, Ethiopia (adapted from Ali, 2005).

	Samp. No	*Temp*	*pH*	*Na^+*	*K^+*	*Ca^{++}*	*Mg^{++}*	*Cl^-*	*HCO_3^-*	*SO_4^{--}*	*F*	*$\delta^{18}O$*	*δD*
1	TD6-1	96	9	620	68	13	0.1	922	45.7	99	1.2	−0.08	−12.00
2	TD6-2	95	9.2	590	68	11	0.1	851	52	109	1.2	−0.11	−13.40
3	Dobi	52	7.4	2125	49.0	4	355	3687	49	505	2.2	−1.16	−8.5
4	Loggya 1	33	8.8	35	4.0	4	26	14	102	57	1.0	−0.29	11.7
5	Loggya 2	33	8.8	35	4.0	4	26	14	102	57	1.0	−0.29	11.7
6	Mille	28	8.1	26	4.0	11	43	12	237	19	0.3	−0.03	12.8
7	Awash R	28	8.8	68	5.0	6	19	31	148	39	0.8	1.65	17.8

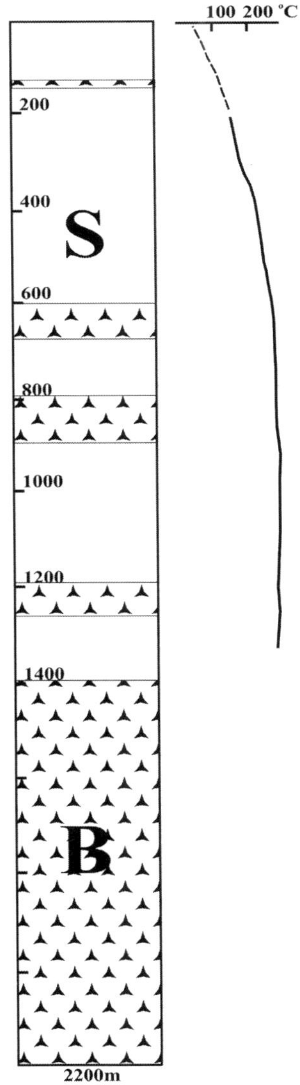

Figure 4.11 Exploratory bore hole in Tendaho. B: Afar Stratoid Series basalts; S: sedimentary rocks such as silt stone, clay-silt mixture and sandstone representing inter-trappeans. The mid hole temperature recorded is about 190°C. Both shallow sedimentary and deep high temperature reservoirs are recognised based on magneto telluric investigations (adapted from Battistellia et al., 2002).

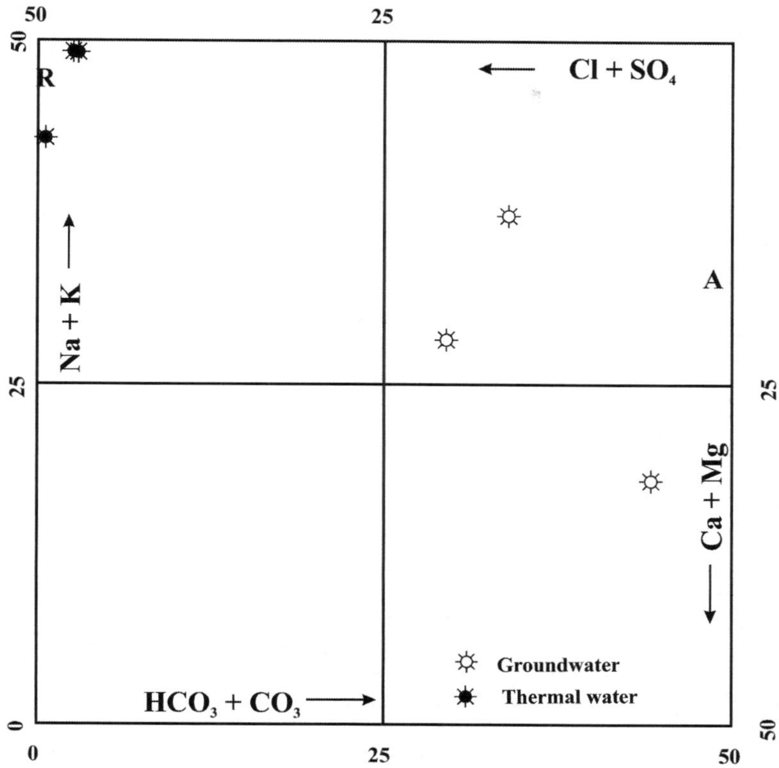

Figure 4.12 Langelier and Ludwig (1942) diagram showing the chemical characteristics of thermal waters and groundwater from the Tendaho geothermal province, Ethiopia (adapted from Chandrasekharam et al., 2015c).

geothermal site and drains into Lake Abhe in Djibouti. Geophysical investigations indicate the presence of a thermal reservoir with a temperature of 200°C (Ali, 2005). A bore hole log from one of the wells drilled in Tendaho shows a very high geothermal gradient in this province; the geothermal reservoir appears to be located in the Stratoid basalts and rhyolites that are found in Djibouti and the northern part of the East African Rift (Figure 4.11).

Magneto telluric investigations have been carried out over the Tendaho geothermal province. The presence of Stratoid Series basalt flows is reflected in high resistivity MT signals at about 2 km depth which is confirmed by an exploratory bore well (Figure 4.11). Low resistivity at 5 km depth is related to the presence of 13% partial melt basaltic magma which confirms the heat source and the presence of a high temperature geothermal reservoir (Didana et al., 2015). Like the geothermal waters in Djibouti and Eritrea, the thermal waters of Tendaho are chloride rich and fall in the chloride field (Figure 4.12) even though the province is away from the Red Sea. Like other west coast Red Sea geothermal provinces described above, the Tendaho thermal systems, originating from a volcanic domain, tend to have high CO_2 content thus

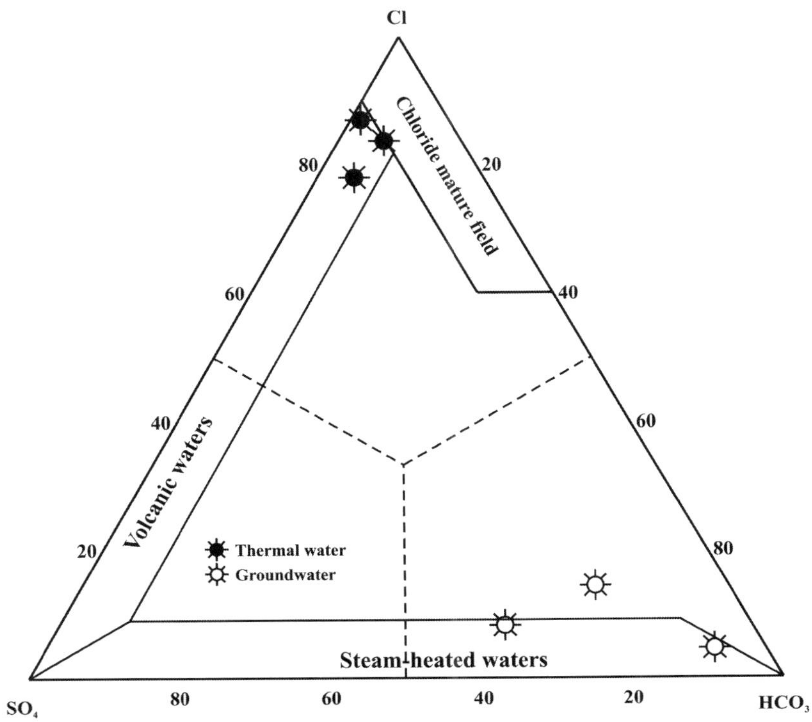

Figure 4.13 Cl-SO_4-HCO_3 Giggenbach diagram (1988) showing the position of Tendaho thermal and cold waters (adapted from Chandrasekharam et al., 2015c).

precipitating carbonate minerals there by enriching the thermal fluids with Na and Cl ions.

The chemistry of the thermal springs corroborates with the geophysical investigation of Didana et al. (2015) and exploratory bore hole records (Figure 4.11) by falling in or near the mature geothermal reservoir field (Giggenbach, 1988). It also indicates a volcanic signature of thermal fluids (Figure 4.13). In the Na-K-Mg diagram of Giggenbach (1988) (Figure 4.14), the thermal waters fall in the partial equilibrium field indicating a prevailing partial equilibrium between the reservoir rocks and the fluids. One thermal water sample from Tendaho falls away from the field indicating a high reservoir temperature that corroborates with the mid hole temperature recorded in the exploratory bore well (Figure 4.11).

The similarity between the geological and tectonic settings, geothermal fluids' characteristics and reservoir characteristics of the Tendaho and Lake Abhe geothermal provinces and the position of these two geothermal systems indicated a similarity between these two provinces in terms of recharge systems and heat sources. Their division into Djibouti and Tendaho is only a political division and has no distinct geographical demarcation. As shown in Figure 4.15, both Tendaho and Lake Djibouti's geothermal systems can be considered as one for developmental purposes.

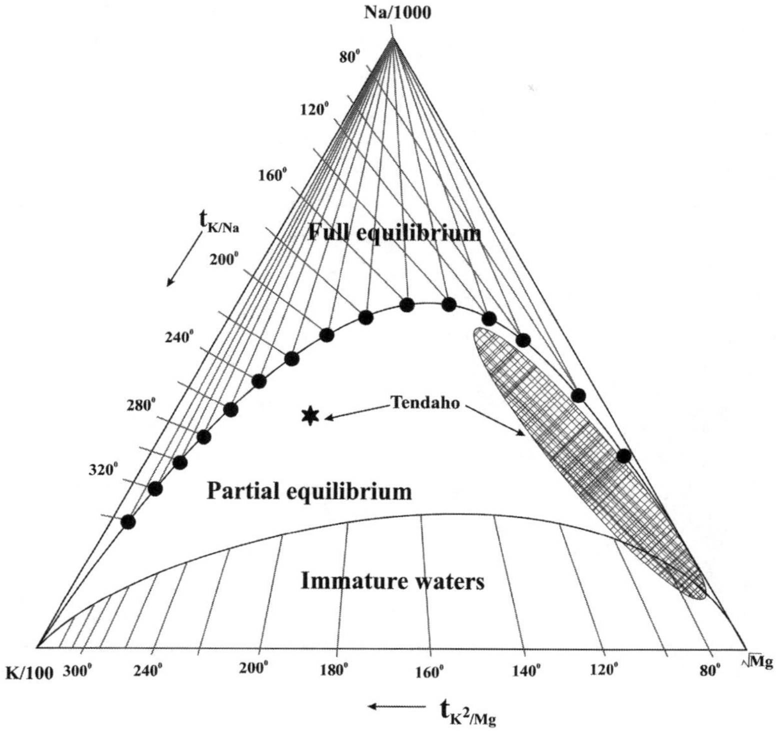

Figure 4.14 Na-K-Mg diagram (Giggenbach, 1988) showing the reservoir temperatures of the Tendaho geothermal province.

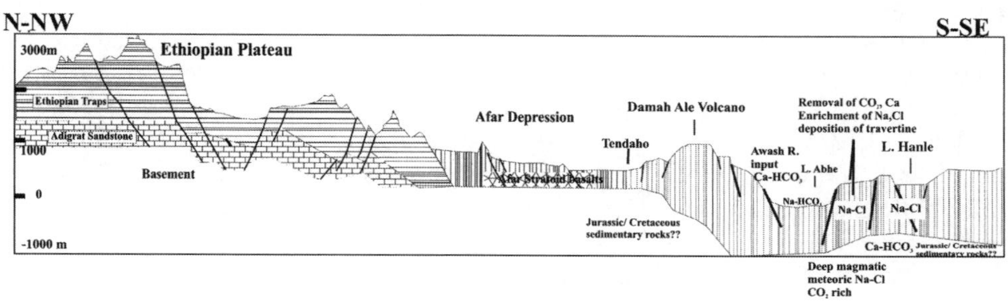

Figure 4.15 Schematic diagram showing the evolution of Tendaho and Djibouti's geothermal systems (adapted from Chandrasekharam et al., 2015c).

4.4 DJIBOUTI

4.4.1 Geological and tectonic settings

The geological evolution of Djibouti is intimately related to the evolution of the Afar, the Gulf of Aden East African Rift, and the Red Sea opening which started around 31 Ma and still continues today. The Djibouti landmass developed over rift formed due to the breakup of the Arabian-Nubian Shield. The remnants of the continental crust are found towards the southern part of Djibouti which is reflected in the Bouguer gravity anomaly. This continental crust attenuates and disappears towards the north (Markis et al., 1975). The oldest volcanic flows over Djibouti, which also covered Ethiopia, are the Adolei basalts that erupted between 27 to 19 Ma (Barberi et al., 1975, Black et al., 1975, Gaulier and Huchin, 1990, Deniel et al., 1994). This was followed by the eruption of the Mabla rhyolites between 16 and 9 Ma (Chessex et al., 1975). This event was succeeded by the eruption of the Dalha basalts between 8 and 6 Ma (Lahitte et al., 2003). This phase was followed by the Stratoid Series, between 4.4 and 0.4 Ma (Barberi et al., 1975). The Stratoid Series flows, erupted due to fissures, cover a large area of 55,000 km^2 with a probable thickness of about 1500 m (Lahitte et al., 2003). This event triggered the opening of the Gulf of Tadjourah (Black et al., 1974) (Figure 4.16). Subsequently, several volcanic eruptions occurred (e.g. Asal series) as a consequence of the intense tectonic activity associated with the opening of the Gulf of Tadjourah and the Gulf of Aden (Fouillac et al., 1989). Sedimentary rocks are present in the coastal areas, in the stream beds and in the south-west tectonic basins. The clays and alluvium deposited during Miocene-Pliocene were interbedded with the Dalha basalt flows. Limestone, clays and diatomites are interbedded within the younger Dalha and Stratoid basalt flows. The thickness of the sedimentary rocks in the plains, exceeds several hundreds of metres while in the stream bed these beds thin out, attaining a thickness of a few metres. Marine sediments (coral limestone and limestone) were deposited in coastal areas during the Quaternary period. Five phases of major extension of lakes were formed during the 100,000 years BP in the tectonic basins when the region experienced a humid climate. During this humid climate period, limestone, diatomites, and clays were deposited (Gasse et al., 1980, CNRS-CNR, 1973).

The most active tectonic region is the Asal rift, located between the Ghoubbet Strait and Lake Asal (Figure 4.16), trending N 135 E. The rift formed about 860,000 years ago accompanied by volcanism giving rise to normal faults. The Moho, in this region, is located at about 6 km beneath the Asal rift (Zlotnicki, 2001). Magneto-telluric data shows the presence of low resistivity zones corresponding to temperatures between 800 to 1200°C. Exploratory bore wells drilled in the Asal rift recorded geothermal gradients of 250°C/km (Zlotnicki, 2001). The western part of Djibouti is also associated with active volcanoes located in Ethiopia (Erta Ale and Damah Ale). Thus, Djibouti, although a small country, is one of the most active regions around the Red Sea that resulted in the evolution of several geothermal provinces around Lake Asal, Lake Abhe, and Lake Hanle (Figure 4.16).

4.4.2 Geothermal manifestations

Djibouti has three well defined geothermal provinces. They are located around Lake Asal, Lake Hanle, and Lake Abhe (Figure 4.16). These provinces are characterised by thermal springs, fumaroles and hot steaming surfaces. The thermal springs in all these three provinces issue through basalt flows (Figure 4.16).

4.4.2.1 Lake Asal

Warm springs with temperatures of 30°C, fumaroles, and hot springs with temperatures of 99°C represent the geothermal setting of Lake Asal (Figures 4.16 and 4.17).

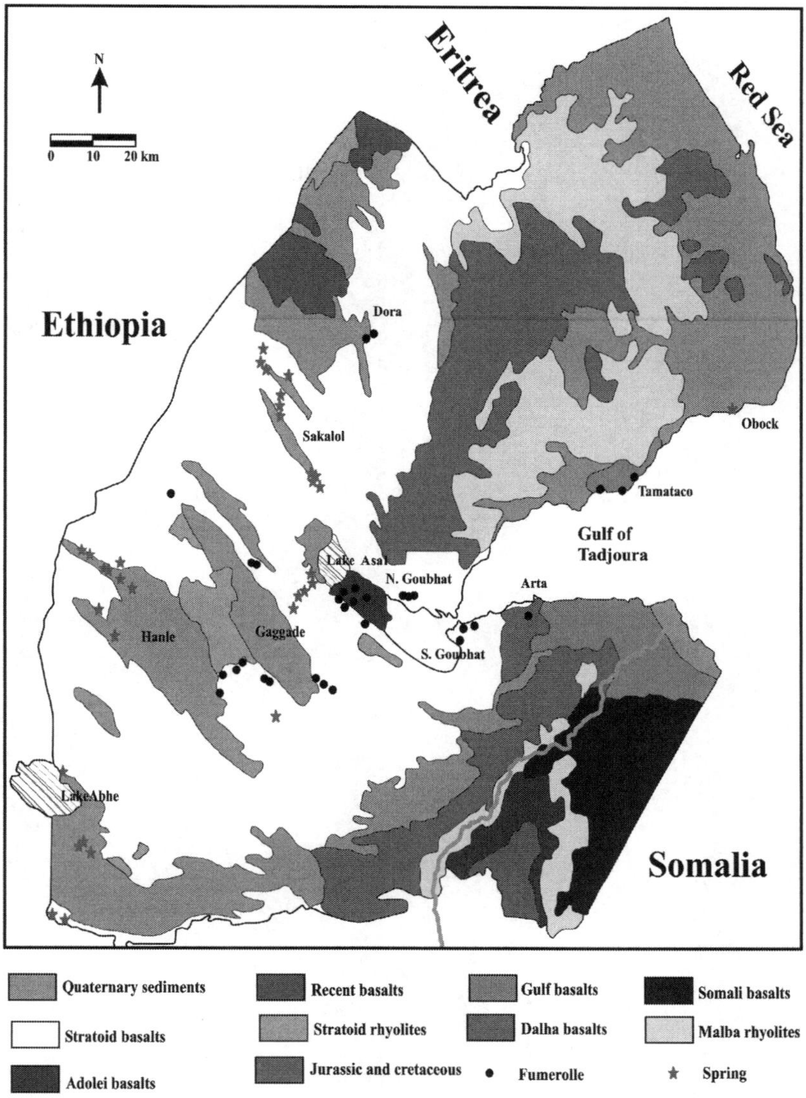

Figure 4.16 Generalised geological map of Djibouti showing the location of the geothermal provinces.

The lake is 150 m below the sea level and sea water is the main source for geothermal systems in this region. Bore hole lithological logs indicate high bottom hole temperatures of 260°C at a depth of about 1500 m (Figure 4.18) (Fouillac et al., 1989, Zan et al., 1990). These warm and hot springs are located along the periphery of the lake and extend towards the southern and northern margins of Goubhat (Figure 4.16).

Table 4.5 Chemical composition of thermal and cold waters from Djibouti (adapted from Boach et al., 1977, Houssein et al., 2013).

	Samp. No	*Temp*	*Area*	*pH*	*Na^+*	*K^+*	*Ca^{++}*	*Mg^{++}*	*Cl^-*	*HCO_3^-*	*SO_4^{--}*	*$\delta^{18}O$*	*δD*
1	2	32	L Assal	7.1	94700	4750.0	2468	11480	183890	125	4300	2.4	1
2	11	32	L Assal	7.0	101100	5160.0	2676	12500	198800	134	4400	−2.1	−0.3
3	7	72	L Assal	6.5	23800	1790.0	7840	1510	58930	31	208	−1.3	−11.7
4	6	83	L Assal	6.1	16300	1380.0	6780	5370	40470	31	126	−1.5	−13.9
5	SP2	96	L Abhe	8.0	1100	31.0	227	0	2072	18	356	−3.48	−23.74
6	SP4	98	L Abhe	8.0	1031	32.0	221	1	2083	16	350	−2.49	−22.18
7	SG3	97	L Abhe	8.4	488	14.0	153	1	987	20	361	−3.42	−24.78
8	SG4	94	L Abhe	8.2	490	14.0	157	0	972	20	356	−3.03	−25.67
9	Niinle	68	L Abhe	7.6	300	21.3	33	3	310	155	200	1.9	−1.3
10	Gorabous	28	L Abhe	7.9	62	5.5	64	22	38	314	75	−0.2	−0.9
11	KK	28	L Abhe	8.3	82	3.9	20	16	44	229	51	−0.69	−1
12	LLB	28	L Abhe	8.6	446	6.3	16	13	262	303	412	−1.16	−0.4
13	R1	28	L Abhe	8.6	501	25.1	29	2	566	146	233	−2.92	−20.4
14	R2	28	L Abhe	8.6	489	18.4	34	1	588	79	247	−2.86	−21.7
15	R3	28	L Abhe	8.6	752	5.5	17	3	700	273	467	−2.23	−17.7
16	Hanle 2	28	L Abhe	8.6	367	7.5	91	1	482	67	334	−3.52	−21.9
17	Lake	29	L Abhe	9.9	43746	435	38	0.05	23462	77763	9768	8.94	47.85

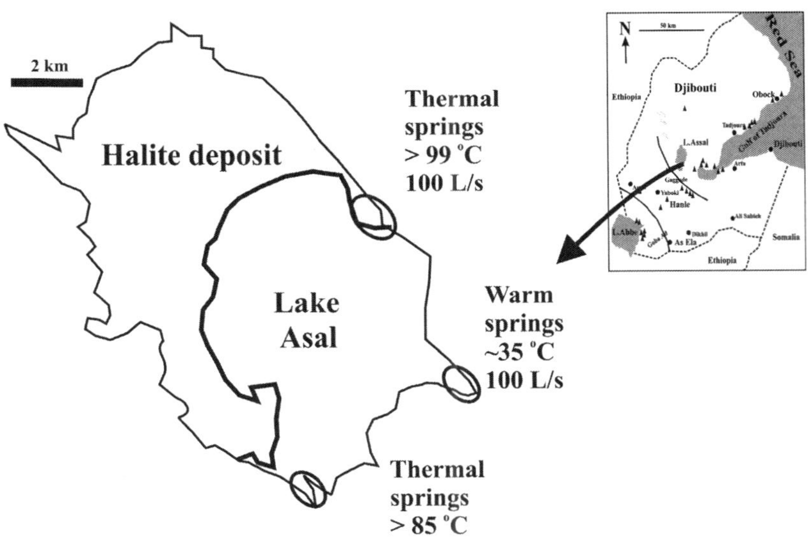

Figure 4.17 Lake Asal geothermal province (adapted from Chandrasekharam et al., 2015c).

4.4.2.2 *Lake Hanle*

Lake Hanle is located between Lake Abhe and Lake Asal in central Djibouti (Figure 4.16) and hosts greater than 13 warm and hot springs. The majority of them flow through Stratoid basalt and rhyolites. Exploratory bore wells were drilled in this region and the lithological sections and bottom hole temperatures measured are shown

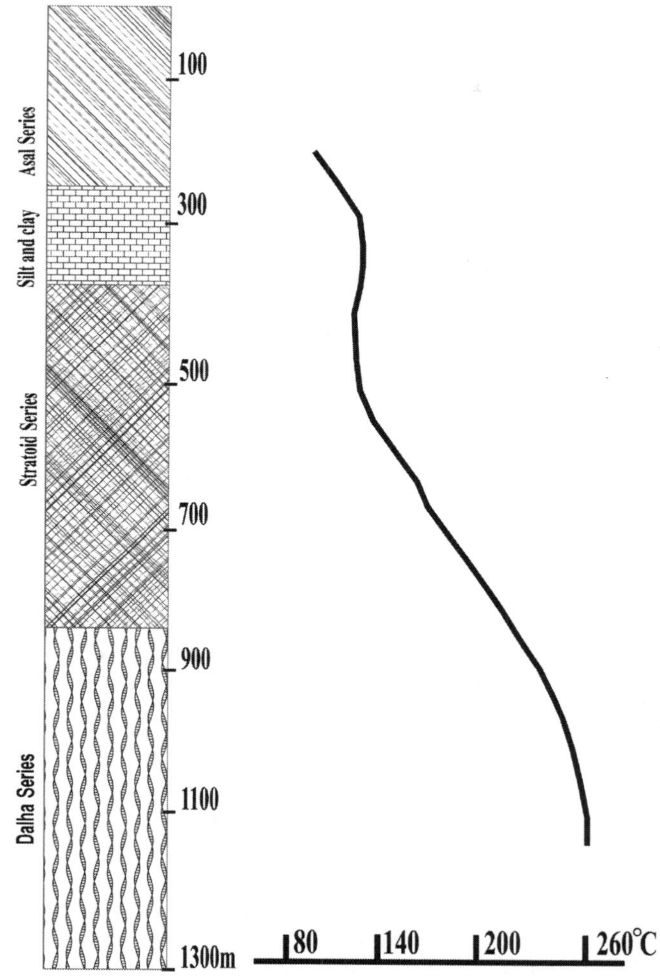

Figure 4.18 Temperature gradient in a bore well in Asal (adapted from Zan et al., 1990).

in Figure 4.19. The bottom hole temperature at 2020 m is less, compared to the bottom hole temperature recorded in the Lake Asal bore well (Figure 4.18). Two wells were drilled in this area to assess the subsurface geology and measure the temperature of the thermal reservoir. These wells penetrated the Stratoid Series basalt flows and the bottom hole temperatures recorded in these wells were 80°C at 1300 m and 124°C at 2020 m (Zan et al., 1990).

4.4.2.3 Lake Abhe

The Lake Abhe geothermal province is located towards the western side of Djibouti at the border between Djibouti and Ethiopia. This province lies east of Damah Ale volcano and south of Erta Ale volcano in Ethiopia. Several thermal springs, fumaroles

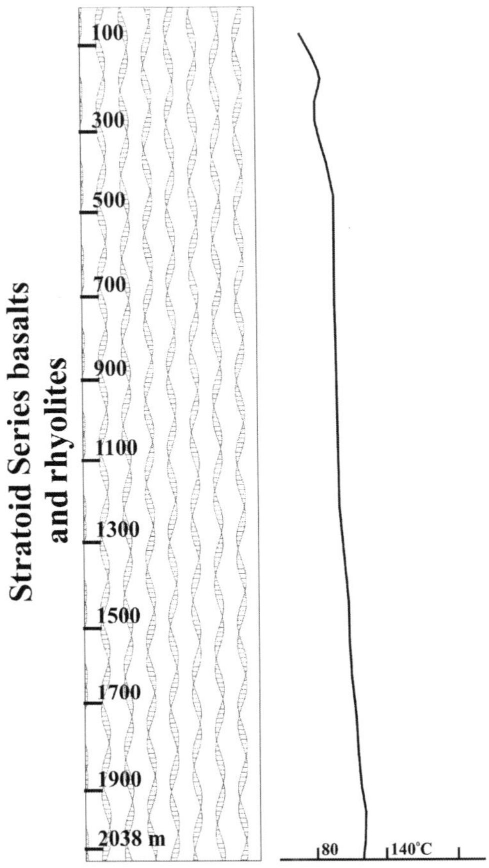

Figure 4.19 Bore hole lithology and bottom hole temperatures, Lake Hanle (adapted from Zan et al., 1990).

and gas vents characterise this province. The temperature of the thermal springs varies from 88 to 98°C and the discharge rate of the thermal waters is about 18 L/s (Chandrasekharam and Chandrasekhar, 2010, Houssein et al., 2013). These springs and gas vents discharge large quantities of CO_2. The release of CO_2 from the linear gas vents and thermal springs resulted in the deposition of travertine which are aligned along linear tracks. The height of these travertine mounds varies from less than a metre to about >15 m. In places, steam discharge can be seen from the travertine mounds.

Lake Asal's thermal and cold springs, due to sea water recharge into the geothermal reservoir, are highly saline and fall in the NaCl field in Figure 4.21. The thermal and cold waters samples from Lake Abhe also fall in the NaCl field even though this region is away from the sea (Figure 4.17). The Lake Abhe water is rich in HCO_3^- and Na^+ and falls in the $NaHCO_3$ field (Figure 4.21). Enrichment in Na^+ in the lake water may be due to precipitation of Ca^{2+} as carbonates in the lake sediments. Although

Figure 4.20 Travertine mounds around Lake Abhe geothermal province (photo by D. Chandrasekharam).

it was not observed, the thermal waters must be discharging into the lake at its bottom. Both Lake Asal and Lake Abhe's geothermal systems are of high temperature and the majority of cold waters show mixing with thermal waters. The Awash River, which drains the Ethiopia rift valley, discharges the sediment and water load into Lake Abhe. Thus, there is a large bicarbonate input into Lake Abhe (Chandrasekharam and Chandrasekhar, 2010, Houssein et al., 2013). The surface manifestation and the temperatures of these two geothermal provinces indicate the presence of high temperature reservoirs in these provinces. This is further supported by the position of the thermal waters in Giggenbach's (1988) Na-K-Mg diagram (Figure 4.22) which indicates reservoir temperatures of 200 to 250°C. The bottom hole temperatures recorded in bore wells in the Lake Asal geothermal site further support the observations in

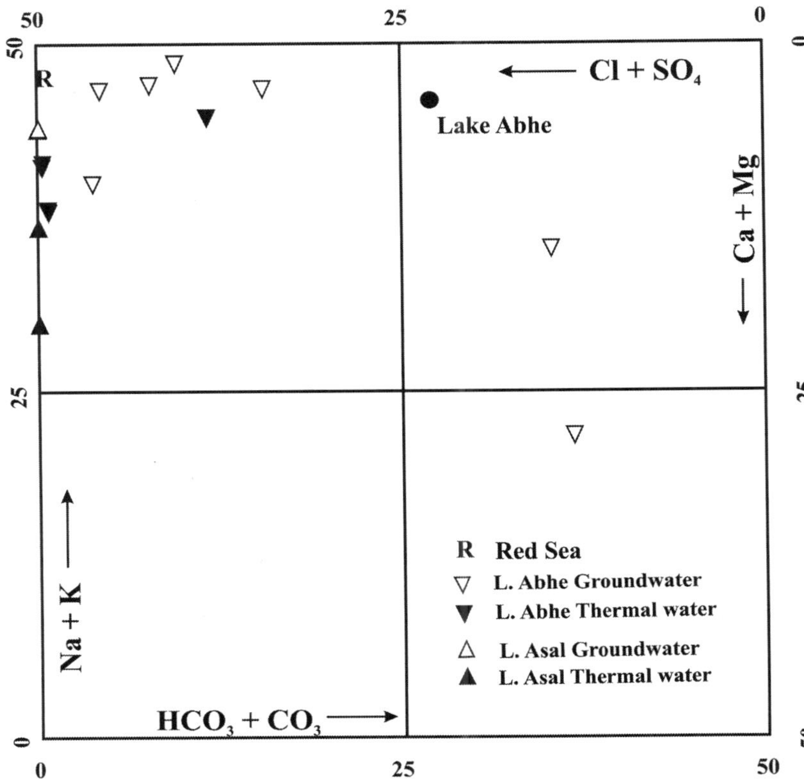

Figure 4.21 Langelier and Ludwig (1942) diagram showing the characteristics of thermal and cold waters from Lake Asal and Lake Abhe geothermal provinces. A Red Sea sample is also plotted for reference (adapted from Chandrasekharam et al., 2015c).

Figure 4.18 and confirm the presence of a high temperature reservoir in this region. Although no bore holes were drilled in the Lake Abhe province, the surface manifestation (fumaroles, gas vents and CO_2 emissions, and travertine mounds) as well as the oxygen isotope values (discussed in the later part of this section) support the presence of a high temperature reservoir in this site also.

4.5 REPUBLIC OF YEMEN

4.5.1 Geology and tectonic setting

The basement rocks in the Republic of Yemen are represented by the Precambrian crystalline and metamorphic rocks that are common in the Western Arabian Shield and Nubian Shield. These rocks are succeeded by Ordovician (quartzitic fluvial deposits), Permian (glaciomarine sediments, shales), Jurassic (Sandstones and shales), Cretaceous (limestone and sandstones) and Quaternary (marine carbonate deposits) formations (Mattash, 1994, Beydoun et al., 1998, As-Saruri, 1999). The basement tectonic fabric

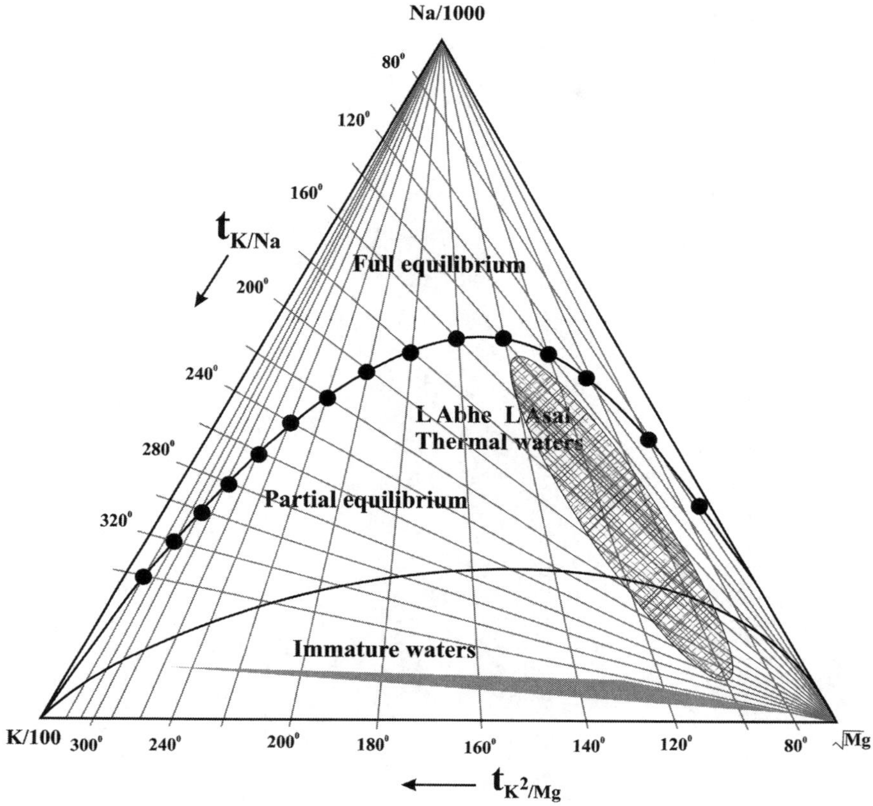

Figure 4.22 Giggenbach's (1988) Na-K-Mg diagram showing the position of the thermal waters from Lakes Asal and Abhe (adapted from Chandrasekharam et al., 2015c).

of the Republic of Yemen follows the regional basement tectonic fabric of the Arabian Shield, represented by the major regional NW-SE trending Najd fault. Magmatic activity prior to the Red Sea rift represented by dike swarms along the western Arabian Shield, is also present in the northern border of the Republic of Yemen, as this part of the landmass is part of the regional Arabian-Nubian Shield (Chandrasekharam et al., 2015c). Volcanic activity over the Republic of Yemen was initiated during Eocene period, with peak activity during Oligocene and Miocene (31–15 Ma). This period coincides with the main rift related volcanic phase over all the land masses around the Red Sea (Plakfer et al., 1987, Menzies et al., 1990, McCombe et al., 1994, Al-Kadasi et al., 1999, Menzies et al., 2001, Minissale et al., 2007, 2013). The volcanic activity continues to be active as evident in the eruption of the Dhamar volcano (Marib) in 1937 and the emergence of two rhyolitic volcanic cones (Quaternary) near Al Lisi and Isbil villages (Figure 4.23). The Oligocene and Miocene volcanism is characterised by a large outpouring of alkaline flood basalts which occupy about 50,000 km^2 in central and western Republic of Yemen. Although volcanic activity during this period is also

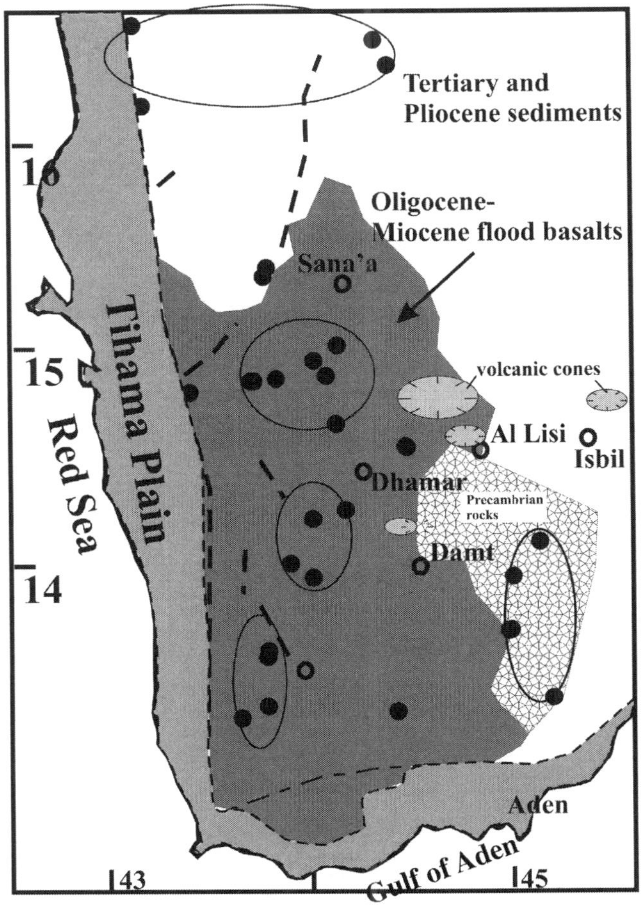

Figure 4.23 Generalised geological and tectonic map of Republic of Yemen showing geothermal provinces (adapted from Chandrasekharam et al., 2015c).

recorded in Eritrea, Ethiopia and Djibouti, on the eastern part of the Red Sea, those volcanic flows are not alkaline.

4.5.2 Geothermal manifestation

More than 100 thermal springs, with issuing temperatures varying from 42 to 96°C, as well as gas vents occur in the western Yemen region (Figure 4.23). Fumaroles and boiling pools are associated with the younger volcanic sites. Most of the thermal springs are fault controlled.

All the geothermal sites are characterised by high heat flow (94 to 154 mWm^2), high geothermal gradients (49 to 77°C/km) and substantial epithermal alterations around the Quaternary volcanic centres (Fara et al., 1999, Minissale et al., 2007, Kohlani, 2013). The presence of these features together with high $He^3/^4He$ ratios (1.2

Table 4.6 Chemical analyses of thermal and cold waters from the Republic of Yemen (adapted from Minissale et al., 2013).

Samp. No	*Temp °C*	*pH*	*Na^+*	*K^+*	*Ca^{++}*	*Mg^{++}*	*Cl^-*	*HCO_3^-*	*SO_4^{--}*	*$\delta^{18}O$*	*δD*
Hamman Ali (Anis)	63	7.9	277	3.6	5	1	83	461	125	−3.67	−19.9
Ad Dunia	65	6.7	135	26.0	202	50	220	232	587	−2.86	−13.9
Kubuth	69	7.6	710	28.0	180	1	1025	58	510	−2.49	−9
Hammam Shra'h	64	7.3	385	16.0	46	9	196	418	438	−3.23	−16.3
Hammam At-Twoair	63	6.9	366	25.0	84	6	350	231	400	−1.66	−3.3
Hammam Al-Sha'rani	70	7.5	125	5.3	11	1	43	204	88	−2.32	−8.3
Hammam Juma'h 2	73	6.9	216	7.8	22	4	135	285	175	−3.32	−18.1
Ma'yoon Al-Haqlah	82	6.9	595	40.0	100	4	925	94	355	−2.13	−11.1
Hamman Ali cold	27	7.7	77	4.3	72	24	49	366	44	−1.75	−7.5
Tuban	32	8.7	274	9.0	78	59	335	171	462	1.27	8.1
Attairi well	21	7.4	43	2.3	80	24	40	378	56	−1	−1.2
Wadi Ad-Durabi	33	7.8	742	37.0	281	10	1020	159	756	−1.11	−3.9
Wadi Mahien	30	8.0	164	2.2	125	44	210	287	188	−0.78	−2.4
Wadi Siham	30	8.1	150	4.0	73	25	119	311	155	−0.66	−1.8
Wadi Melah	31	7.8	182	3.1	85	5	150	210	140	−3.02	−16.2
Wadi Surdod	32	8.4	51	4.2	47	16	52	195	75	−1.12	−1.6
Wadi Ar-Raboa'	29	7.8	40	1.3	75	16	105	240	28	0.1	5.5
Wadi Dyyan	28	8.3	49	2.3	69	19	43	275	60	−0.18	1.1
Al-Ain Albared	27	7.8	10	1.4	47	12	14	164	24	−1.9	−3.8
Wadi Habel	32	8.7	226	5.0	61	46	315	232	219	0.12	0.4
Al-Jammah Wadi Al-Meer	28	8.7	37	1.4	35	10	24	183	23	0.52	8.1
Wadi Laa'h	35	8.7	60	4.1	35	19	86	92	119	1.41	11.5

to 3.2, Minissale et al., 2007) in the thermal gases indicate the presence of a shallow magma chamber within the crust in this region (Dowgiallo, 1986, Fara et al., 1999, Mattash et al., 2005). The chemical composition of certain thermal springs and groundwaters from the Republic of Yemen is shown in Table 4.6.

In the Langelier and Ludwig (1942) diagram (Figure 4.24), the thermal and cold water samples show wide compositional variation. The thermal waters fall in the NaCl and $NaHCO_3$ fields while some of the groundwater samples show typical $CaHCO_3$ character and others show a mixing relationship, probably with the saline thermal waters. The thermal gases are rich in CO_2 and the thermal waters registered very high pCO_2 (0–0.25) and high $\delta^{13}C$ (−4 to −8‰) thereby indicating a mantle signature (Minissale et al., 2007). Due to the high concentration of CO_2 in the thermal gases, precipitation carbonate minerals as travertine deposits around the thermal provinces have been reported (Minissale et al., 2007). The thermal waters are also characterised by high fluoride (1.8 to 14.5 ppm, Minissale et al., 2007, 2013) and bicarbonate content, similar to those reported in thermal waters from the East African Rift valley (Gizaw, 1996). Geothermal systems associated with volcanism tend to contain high CO_2 partial pressures thereby precipitating $CaCO_3$ and leaving the thermal fluids rich in $NaHCO_3$ and fluoride (Gizaw, 1996).

Thus, even though some thermal waters fall in the bicarbonate field in the Giggenbach's (1988) Cl-SO_4-HCO_3 diagram (Figure 4.25), this is mainly due to the precipitation of calcium carbonate minerals and not due to the dilution or mixing with

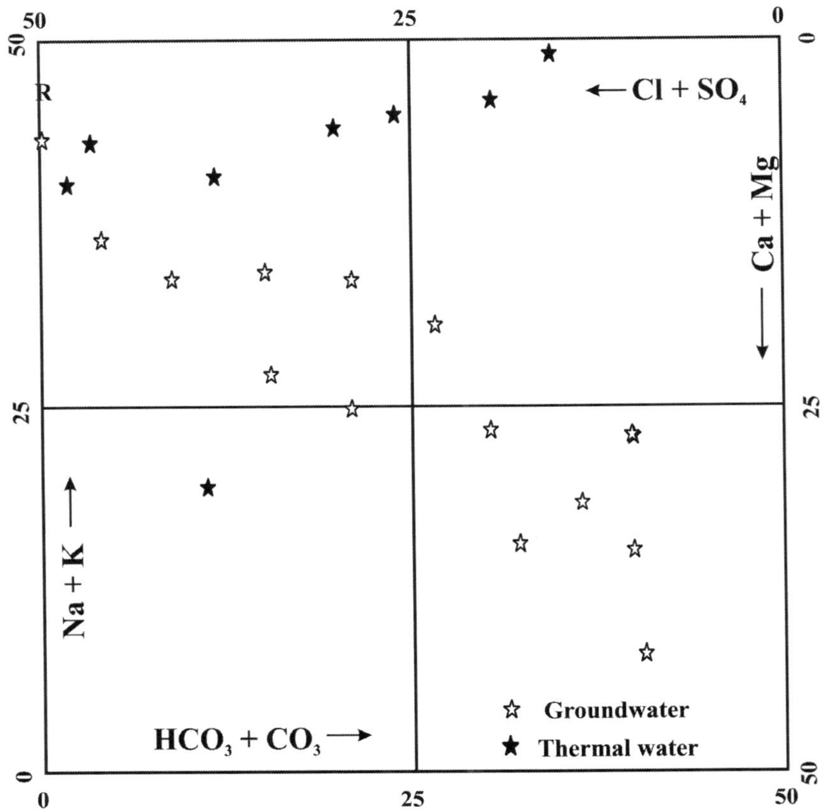

Figure 4.24 Langelier and Ludwig (1942) showing the chemical characteristics of thermal and cold waters from the Republic of Yemen. Red Sea sample is also plotted for reference (adapted from Chandrasekharam et al., 2015c).

near surface groundwater. The plots of other thermal waters in Figure 4.25 clearly demonstrate the influence of the volcanic gases on the circulating thermal fluids in Yemen. The presence of high temperature geothermal systems in this region is further indicated in the Na-K-Mg diagram of Giggenbach (1988) where most of the samples plot in the partial equilibrium field and those falling in the immature field may either indicate a mixing with the HCO_3 rich groundwaters or an enrichment of Mg relative to a precipitation of $CaHCO_3$ as travertine (Figure 4.26).

4.6 SAUDI ARABIA

4.6.1 Geology and tectonic setting

The western Arabian Shield is part of the Arabian-Nubian Shield which experienced a series of tectonic, magmatic and metamorphic events since the Precambrian. The

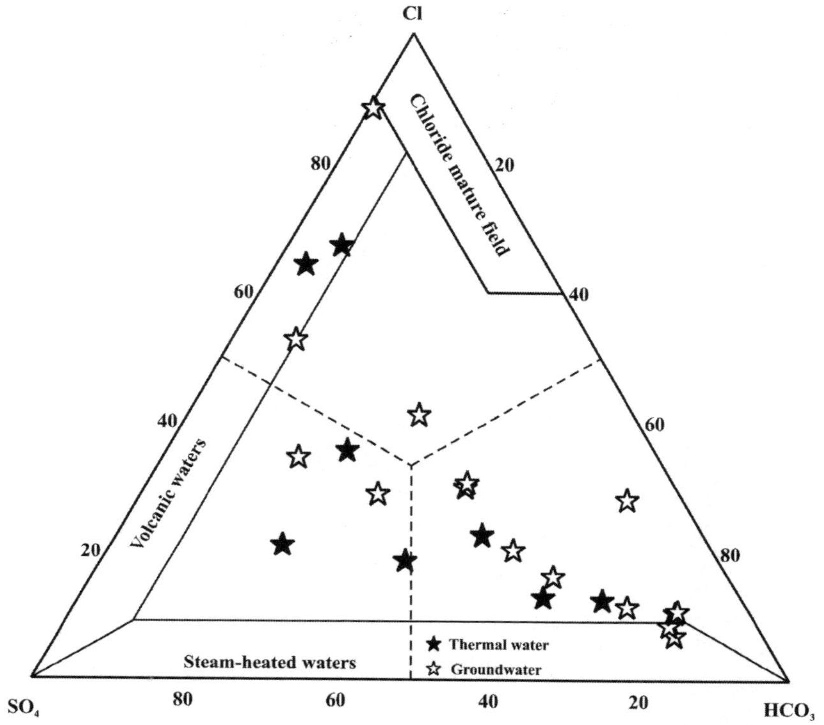

Figure 4.25 Cl-SO_4-HCO_3 diagram showing the volcanic signature in the Republic of Yemen thermal waters.

tectonic activity was intense between 900 and 550 Ma (Stern, 1994). These events are not local or confined to the Arabian Shield, but are regional. Kennedy (1964) termed these events as an orogenic cycle known as the Pan-African Orogeny (PAO) and the Pan-African Thermo-Tectonic episode. Kroner (1984) states that the events and style of tectonism is common to all of Gondwanaland, the time when all the landmasses were juxtaposed (Figure 4.27).

The ANS is a conglomeration of older crustal rocks and Neoproterozoic terranes caught in between East- and West Gondwanaland (Figure 4.27, Johnson and Woldehaimanot, 2003). While the Precambrian rocks are exposed on either side of the Red Sea, Neoproterozoic terranes are preserved in the eastern side of the Red Sea over the Arabian Shield. A large part of the Nubian Shield on the western part of the Red Sea is represented in Egypt, Eritrea and part of the East African Plate. The Arabian Shield, as the name defines, occupied a large part of the western part of Saudi Arabia on the eastern side of the Red Sea, after it was separated and rotated due to the opening of the Red Sea. The formation of young rifts, narrow seas parallel to the coasts, development of ocean basins associated with mid ocean rifts, evolution of island arcs and trenches, compression of older land masses to give rise to folded mountain belts sandwiching ancient crustal rocks and basic volcanism associated with younger tectonic processes

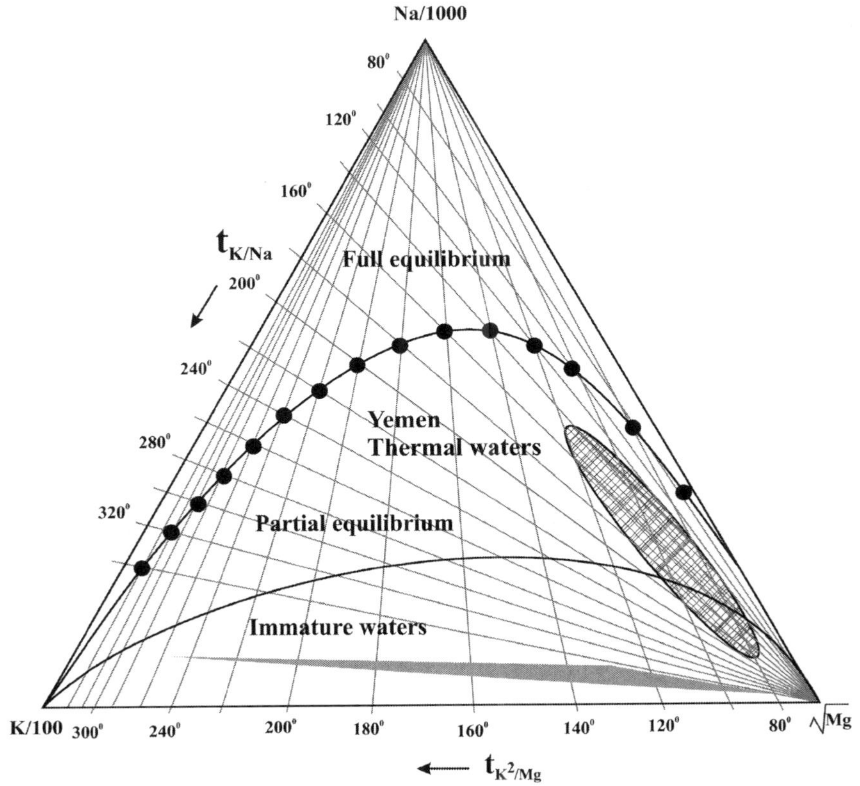

Figure 4.26 Na-K-Mg diagram (Giggenbach, 1988) showing the high temperature characteristics of the Republic of Yemen's geothermal system.

are all part of the Wilson cycle, named after J. Tuzo Wilson who recognised this cycle in 1965 (Wilson, 1968) and the entire cycle is generally represented in a "Shield". The PAO tectonics and orogenic cycle are entirely different from what prevailed during the period between 1.8 and 1 Ga. Prior to the tectonic and magmatic events that reshaped the ANS between 900 and 400 Ma and later by younger magmatic and magmatic and tectonic events after 30 Ma, the ANS is considered as a juvenile crust evolved between the east and west Gondwana (Figure 4.48), were amalgamated pieces of land and interwoven with several arcs and subduction tectonics represented by ophiolite belts that are exposed over the present day shield regions of Arabia and western part of the Red Sea.

By late Proterozoic, the Arabian Shield had developed into distinct terranes, separated by ophiolite bearing suture zones. Three ensimatic island arc type terranes developed in the western shield while two terranes with continental affinities were formed on the eastern side of the shield (Figure 4.29). The accretion of these terrains represents the present day western Arabian Shield (Stoeser and Camp, 2014).

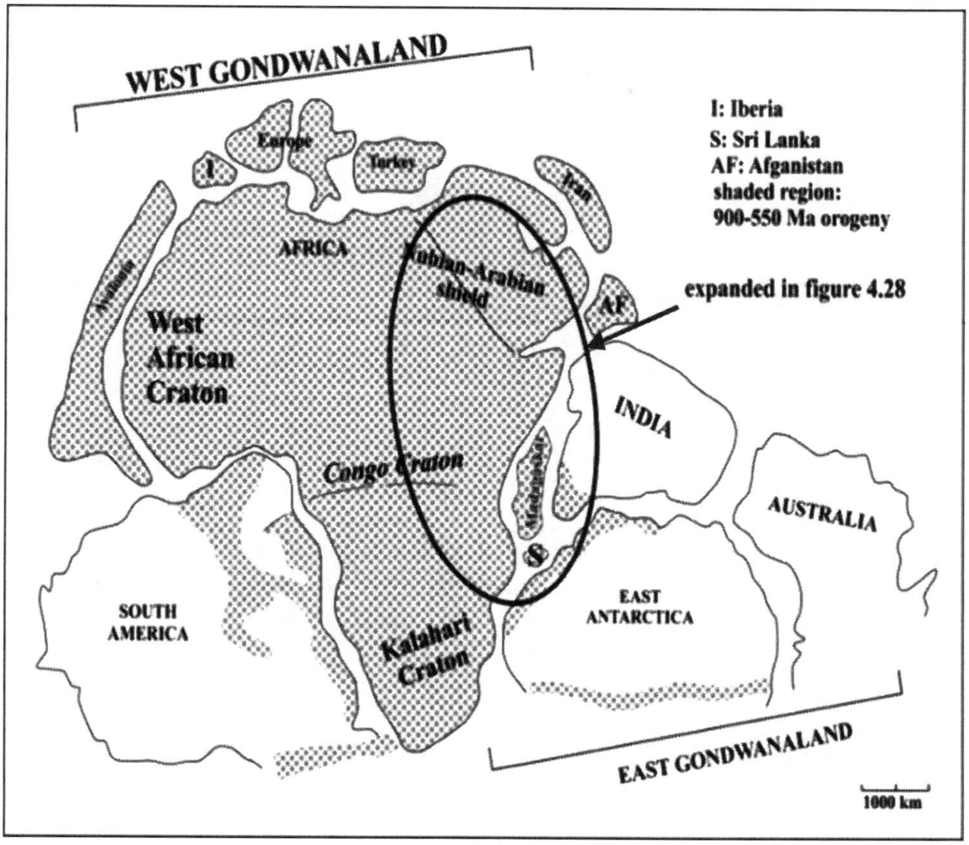

Figure 4.27 Configuration of landmasses prior to the onset of the Red Sea rift and breakup of Nubian Shield. The shaded area shows the regions affected by the Pan-African Orogeny (adapted from Stern, 1994).

Although the shield has been subjected to several phases of tectonic and magmatic events between 500 Ma and Recent, these suture zones with ophiolites are still preserved and host several minerals and metals of economic value (Al-Shanti and Roobol, 1979, Ahmed and Habtoor, 2015). Between 900 and 680 Ma, the western Arabian Shield experienced intense plutonic activity, dominated by intermediate plutonic rocks such as diorite, tonalite and trondhjemite, and the evolution of primitive tholeiitic, and calc-alkaline series of rocks evolved under an ensimatic and continental margin arc environment. Between 680 and 631 Ma collision tectonics prevailed between the accreted ensimatic arc terrane and continental plate and between two continental plates. At this stage, due to collision tectonics, the magmatic style changed from an arc to a collision type magma generation and magmatic activity was at its peak between 660 and 610 Ma. This magmatic phase is represented by an evolution of granites and related rocks. Between 610 and 510 Ma intracratonic magmatism was

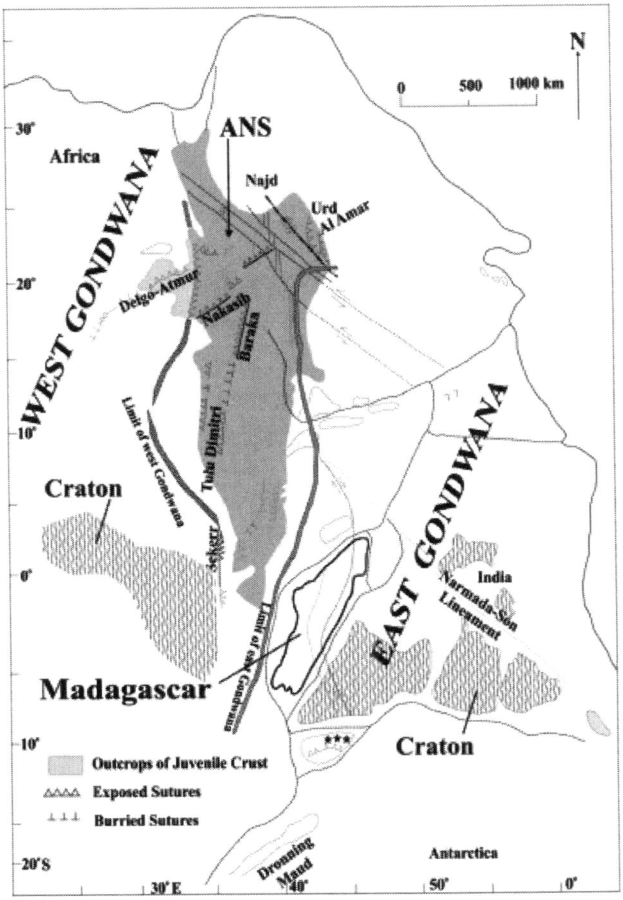

Figure 4.28 The disposition of juvenile land masses between east and west Gondwana, with suture zones and ophiolite belts and the evolution of ANS. Remnants of these sutures became part of the Arabian Shield after the opening of the Red Sea surrounded by terranes (adapted from Stern, 1994, Johnson and Woldehaimanot, 2003, Kroner and Stern, 2004).

prevalent giving rise to peralkaline alkali feldspar granites and peraluminous granites and syenites. These granites typically represent the end phase of the continental collision (Stoeser, 1986). The percentage of granites and related plutonic rocks out cropping in different terrane is given in Table 4.7 and their distribution is shown in Figure 4.30. These granites and related rocks occupy an area of about 161,467 km^2 in the shield (Chandrasekharam et al., 2014, Lashin et al., 2014a,b).

The most significant feature of these post orogenic plutonic rocks is that they contain a high content of uranium, thorium and potassium and hence generate a large amount of heat (6 to 134 μWm^{-3}) and therefore the heat flow value over the terranes containing these rocks is also very high (>80 mW/m^2) (Chandrasekharam et al., 2014b, 2015d). Because of this extraordinary characteristic, these rocks are able to host high

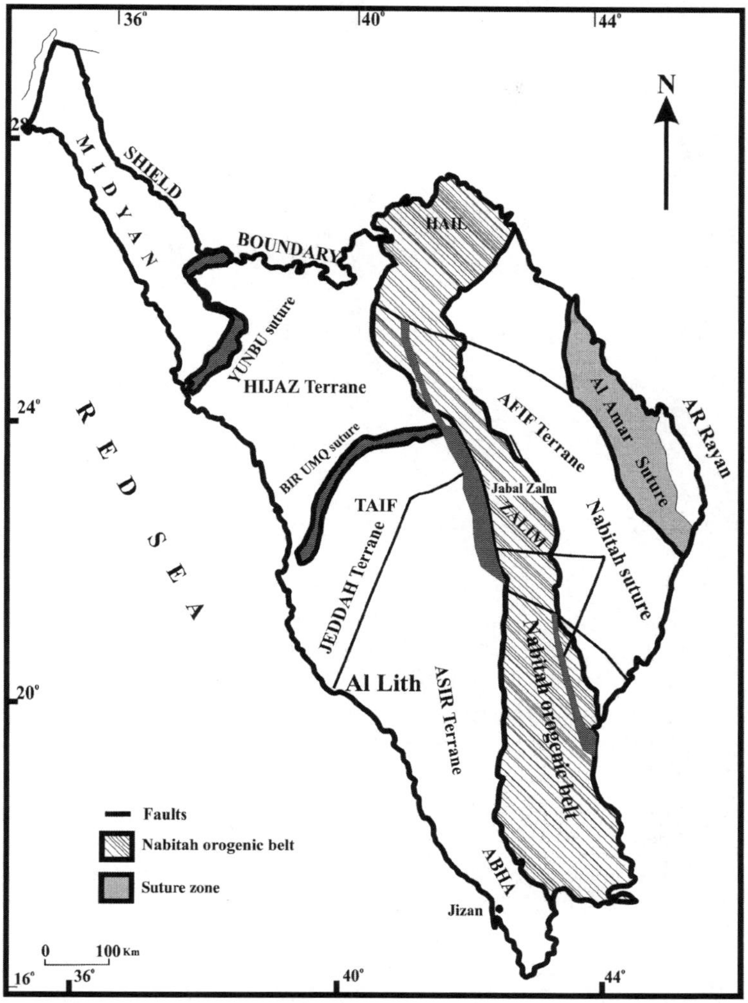

Figure 4.29 Sutures and terranes of the Arabian Shield (adapted from Stoeser, 1986).

temperature geothermal systems along the western coast of the shield. Chapter 9 in this book is entirely devoted to such rocks from the continents around the Red Sea. Therefore, further details are not provided here in this chapter.

After the post orogenic events, the ANS experienced a major tectonic and magmatic episode between 31 and 5 Ma which triggered the Red Sea rift process that resulted in the formation of the Red Sea and the breakup of the ANS. Since these geological processes are responsible for the evolution of all the landmasses and associated geothermal systems around the Red Sea, a detailed account of these magmatic and tectonic processes are described in detail in section 4.7 below. A short account of the evolution of volcanism over the Arabian Shield due to the initiation of the Red Sea rift

Table 4.7 Outcrop of post orogenic felsic and mafic plutonic rocks (%) in the five terranes (adapted from Stoeser, 1986 and Chandrasekharam et al., 2014).

	M	*H*	*A*	*Af*
Alkali felspar granite	14	10	3	4
Granite	45	31	57	48
Granodiorite	20	33	71	36
Tonolitic rocks	15	15	52	8
Gabbroic rocks	6	9	16	3
Svenitic rocks	0	2	1	1
Total granite rocks	79	74	91	88
Total intermediate rocks	21	26	9	12

M: Midyan terrane; H: Hijaz terrane; A: Asir terrane; Af: Afif and Ar Raynn terranes.

process that encompasses the origin of geothermal systems over the Arabian Shield is given here.

The genesis of the hydrothermal systems is coeval with the post orogenic magmatism and the breakup of the ANS between 31 and 5 Ma. The initial breakup was triggered by the Afar plume which covered a large part of the ANS. The plume head, according to seismic tomographic analysis (Debayle et al., 2001), was positioned below Ethiopia and its periphery extended to a large area covering the southern part of Saudi Arabia, the Republic of Yemen, Djibouti, Eritrea and Egypt, land masses with similar tectono-magmatic and geothermal settings (Figure 4.31).

The initial plume propelled the Red Sea rift which started from the southern part (Present Aden-Afar-Red Sea axis junction) and propagated northwards. This process started around 31 Ma and the volcanic activity is still continuing over the Arabian Shield and Eritrea (Hamlyn et al., 2014, Moufti et al., 2013, Duncan and Al-Amri, 2013). As a consequence of this rift propagation, the western coast of the shield experienced regional dike swam activity parallel to the Red Sea rift axis (Bayer et al., 1989, Camp and Roobol, 1992, Bosworth et al., 2005) and this process is still active as evident from the recent earthquake swarm below Harrat Lunayyir (Figure 4.32) (Pallister et al., 2010, Al-Shanti and Mitchell, 1976, Duncan and Al-Amri, 2013).

Post rift tectonic activity resulted in the eruption of large volumes of volcanic flows over the entire Arabian Shield and the Republic of Yemen. (Figure 4.32). and occupy and area of about 90,000 km^2 (Coleman et al., 1983). A few of them, as stated above, are still active, with fumaroles and hot springs. The basalt flows covered a large part of the paleo channels giving rise to hot aquifers below the harrats. The steam from these aquifers, and the steam separated from the basaltic magma located within crustal levels, have given rise to fumaroles around several harrats. These areas record a high geothermal gradient (90°C/km, Coleman et al., 1983) and hence a high heat flow. Many of the post orogenic granites outcropping along the coast are covered by the younger volcanic flows (Figure 4.30). These volcanic centres and the high heat generating granites host high temperature geothermal systems (Chandrasekharam et al., 2014b, 2015b,c, Lashin et al., 2014a).

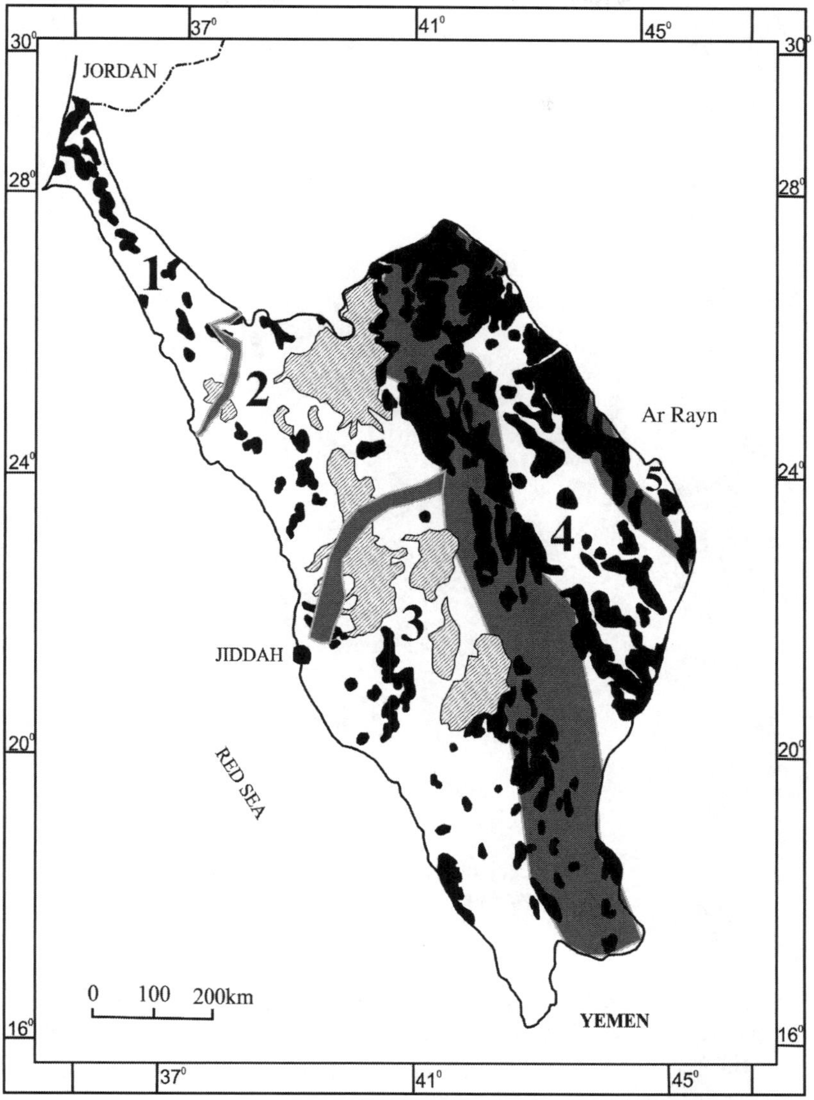

Figure 4.30 Distribution of plutonic rocks in the terranes, Arabian Shield 1. Midyan, 2. Hijaz, 3. Asir, 4. Afif, and 5. Ar Rayn. Dark grey: Suture, light grey: Harrats, black: felsic intrusives (adapted from Stoeser, 1986 and Chandrasekharam et al., 2014b).

4.6.2 Geothermal systems

The western Arabian Shield is the locus of high temperature geothermal systems represented by the emergence of more than 50 thermal springs with surface temperatures varying between 31 and 96°C with flow rates of 5 to 20 L/s (Al-Dayel, 1988, Lashin and Al Arifi, 2012, Chandrasekharam et al., 2014a,b, 2015b,c, Lashin et al., 2014,

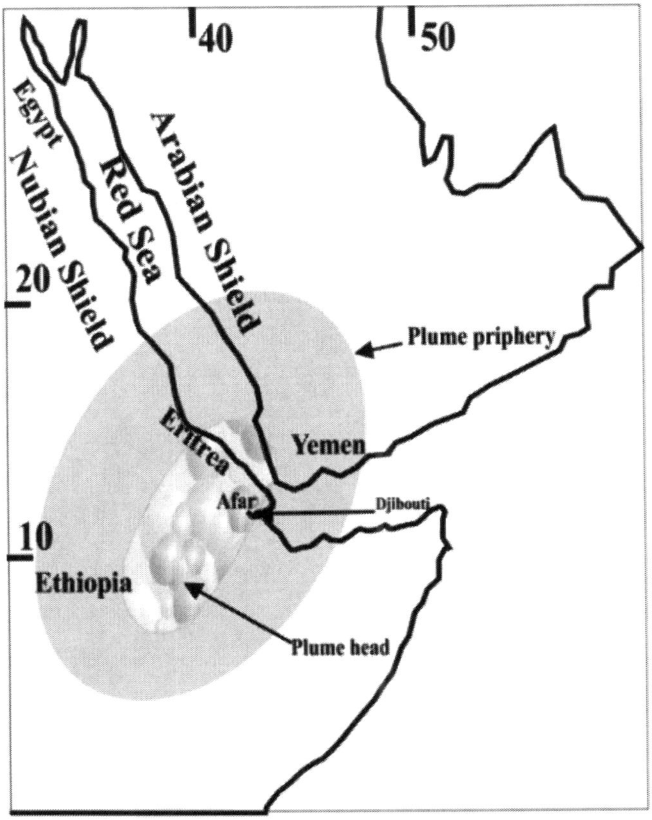

Figure 4.31 Influence of the Ethiopian plume over the continent around the Red Sea (adapted from Debayle et al., 2001, Bosworth et al., 2005).

2015). These springs can be classified into two groups: 1) those associated with volcanic centres (harrats) and 2) those associated with high heat generating granites. The locations of the geothermal sites are shown in Figure 4.33.

Those systems that are associated with the harrats are high temperature fumaroles described in section 4 above. These steam pools are generated due to the escape of steam from rising magma, or degassing in the magma chamber, or due to the boiling aquifers (paleo channels) that are buried by the volcanic flows in all the harrats. These geothermal systems are similar to those operating in Ethiopia (Tendaho), Djibouti (Lake Abhe), Eritrea (Alid and Nabro-Dubbi) and the Republic of Yemen (SW-W and N of Damar). These thermal springs are fault controlled with regional faults running parallel to the coast and are aligned with the major Najd fault system of the Arabian Shield. The evolution of these geothermal systems is coeval to the evolution of the harrats and associated basic intrusives (dike swarms) along the coast and continue to be active due to the prevailing high heat flow and high geothermal gradients caused by

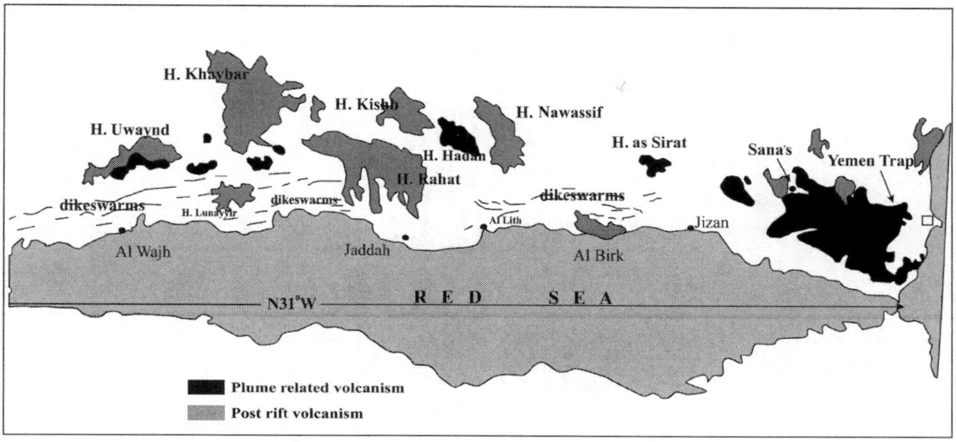

Figure 4.32 Plume related and post rift volcanism over the Arabian Shield and the Republic of Yemen (Bosworth et al., 2005).

intrusion of magma at shallower levels and conduction of heat from the paleo suture zones and shallow Moho (~20 km, Windley et al., 1996, Bohannon, 1987, Bayer et al., 1989, Al-Dayel, 1988, Koulakov et al., 2014, 2015). The heat flow values recorded over the shield region around the geothermal provinces is >80 mW/m^2 (Gettings et al., 1986) while the normal heat flow values over the shield are 50 and above mW/m^2 (Rybach, 1976). The chemical analyses of thermal, warm and cold waters from selected sites in the shield region are shown in Table 4.8.

The thermal, warm and cold water samples form an array along the Na+K axis (Figure 4.34) and fall within the Cl+SO_4 field indicating enrichment of Cl^- over HCO_3^-. Even the warm wells and cold waters also have similar characteristics. Although the water samples are rich in Cl^-, the Red Sea is located away from the site of the springs and the Red Sea sample is distinctly different from these saline waters (Figure 4.34). The Cl/HCO_3 ratio and SO_4/HCO_3 ratios of the Red Sea are too low compared to all the water samples shown in Table 4.8 and a strong inter-ion relationship that generally exists in sea water is lacking in the geothermal waters from this region (Lashin et al., 2014). The groundwater that occurs in shallow aquifers is HCO_3 rich and generally plots in the HCO_3 field. In the present case both warm and cold waters appear to have a genetic relationship. In several geothermal systems this phenomenon is common due to the mixing of ascending thermal water with the near surface groundwater resulting in an enrichment with Cl in the mixed water. The most interesting aspects of the thermal waters is high fluoride content (Table 4.8). This indicates that the thermal fluids are interacting with rocks with high fluorine content either in the reservoir or during their ascent. This is ascertained by the presence of high fluorine bearing minerals like fluorite and apatite in the felsic rocks. The fluorine content in these rocks varies from 50 to >3450 ppm (du Bray et al., 1983, Stoeser, 1986, Agar, 1992, Kuster, 2009).

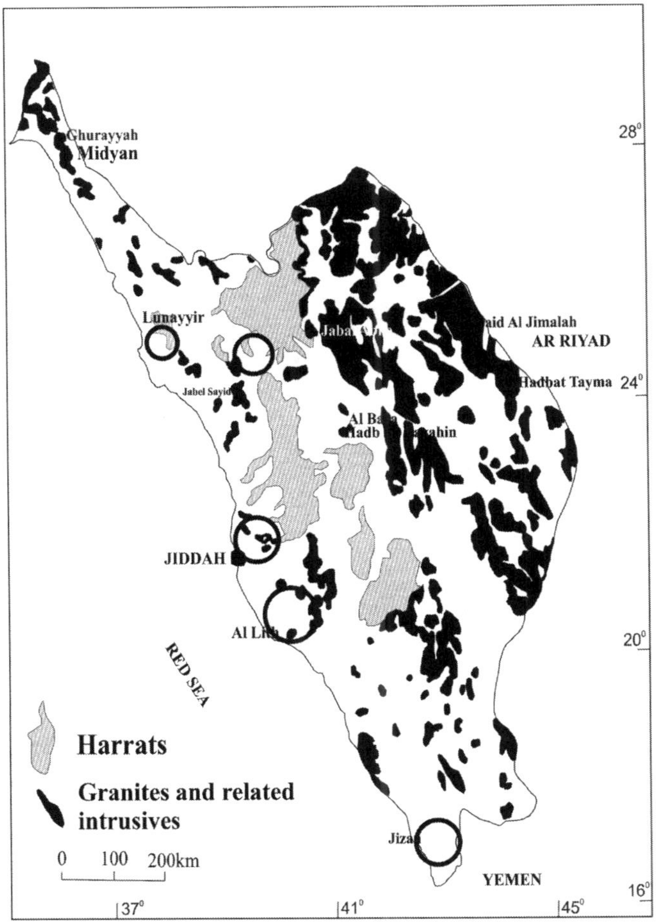

Figure 4.33 Western Arabian shield showing the geothermal provinces along the Red Sea coast. These provinces are associated with harrats and granites (adapted from Chandrasekharam et al., 2014a).

The thermal waters have attained partial equilibrium with the reservoir rocks, indicating reservoir temperatures between 120 and 160°C, while the warm springs that evolved due to the mixing of shallow groundwater, fall in the immature field (Figure 4.36).

4.7 OXYGEN AND HYDROGEN ISOTOPE BEHAVIOUR OF THE THERMAL SPRINGS

The oxygen and hydrogen isotope ratios of thermal waters from all the countries around the Red Sea discussed above are plotted on the $\delta^{18}O$ *vs* δD diagram (Figure 4.37)

Table 4.8 Chemical characteristics of thermal, warm and cold waters from the Arabian geothermal provinces (source Lashin et al., 2015, Chandrasekharam et al., 2015b,c).

	Locality	*Site*	*Type*	*Temp* °C	*pH*	Na^+	K^+	Ca^{2+}	Mg^{2+}	SO_4^{2-}	HCO_3^-	Cl^-	F^-	$d^{18}O$	*dD*	P_{CO_2}
1	Jizan	Ain Al Wagarah-1	HS	44	7.7	343	50	367	20	390	24	1083	3.1	−3.66	−14.8	−3.32
2	Jizan	Ain Al Wagarah-2	HS	45	7.5	621	111	601	57	415	22	2059	3.1	nd	nd	−3.18
3	Jizan	Ain Al Wagarah-3	HS	57	7.2	317	40	326	32	260	22	923	2.6	nd	nd	−2.73
4	Jizan	Ain Al Wagarah-4	HS	57	7.2	339	46	317	31	208	18	1243	3.3	nd	nd	−2.82
5	Jizan	Ain Al Wagarah-5	HS	45	7.2	134	11	329	21	248	22	959	2.8	nd	nd	−2.8
6	Jizan	Ain Al Wagarah-6	HS	61	7.0	533	40	341	15	215	20	941	3.2	nd	nd	−2.54
7	Jizan	Ain Al Wagarah-7	HS	57	7.6	199	6	302	20	193	12	504	3.2	nd	nd	−3.43
8	Jizan	Ain Khulab	HS	76	7.4	473	24	429	60	238	16	586	2.2	−3.16	−21.3	−3
9	Jizan	Dani Malik-2	HS	45	7.7	210	12	270	7	235	31	309	3.0	nd	nd	−3.16
10	Jizan	Al Ardah-3	WS	31	8.0	12	3	159	75	58	30	64	0.5	nd	nd	−3.56
11	Jizan	Al-Ardah-6	WS	33	7.0	24	6	1631	158	440	45	1321	0.7	nd	nd	−2.47
12	Jizan	Al Khouba-2	WS	33	7.4	45	9	248	39	160	36	75	0.4	nd	nd	−2.85
13	Jizan	NE Jazan-22	WS	32	7.2	406	5	139	93	485	94	2077	0.8	nd	nd	−2.25
14	Jizan	NE Jazan-23	WS	30	6.9	321	4	124	69	525	94	495	0.8	nd	nd	−1.95
15	Jizan	NE Jazan-24	WS	33	6.9	356	4	129	74	453	94	2126	1.0	nd	nd	−1.94
16	Jizan	NE Jazan-29	WS	32	7.1	247	3	178	93	686	119	1879	nd	nd	nd	−2.05
17	Jizan	NE Jazan-9	WS	34	6.9	593	7	228	157	1175	291	247	nd	nd	nd	−1.45
18	Al Lith	92	HS	41	7.4	249	24	228	3	243	21	865	2.3	−3.7	−19.7	
19	Al Lith	102a	HS	56	6.9	594	30	541	5	450	18	1584	3.2	−4.2	−23.1	
20	Al Lith	110a	HS	41	7.6	484	29	684	3	575	16	760	2.4	−4.2	−23.2	
21	Al Lith	112a	HS	96	7.8	404	25	260	0	225	18	845	2.5	−3.8	−21.3	
22	Al Lith	113a	well	34	7.5	213	9	103	38	340	39	186	0.6	nd	nd	
23	Al Lith	114a	well	33	8.4	179	10	118	37	315	32	126	0.4	nd	nd	
24	Al Lith		HS	59	7.9	305	17	94	5	430	107	270	nd	−2	−4	
25	Al Lith		HS	59	7.5	492	28	116	19	600	70	477	nd	−3.3	−14.3	
26	Al Lith		HS	79	7.7	424	12	234	0	440	21	687	nd	−3.8	−21.3	
27		Red Sea			8.2	12600	843	901	1641	13400	26772	26354	nd			

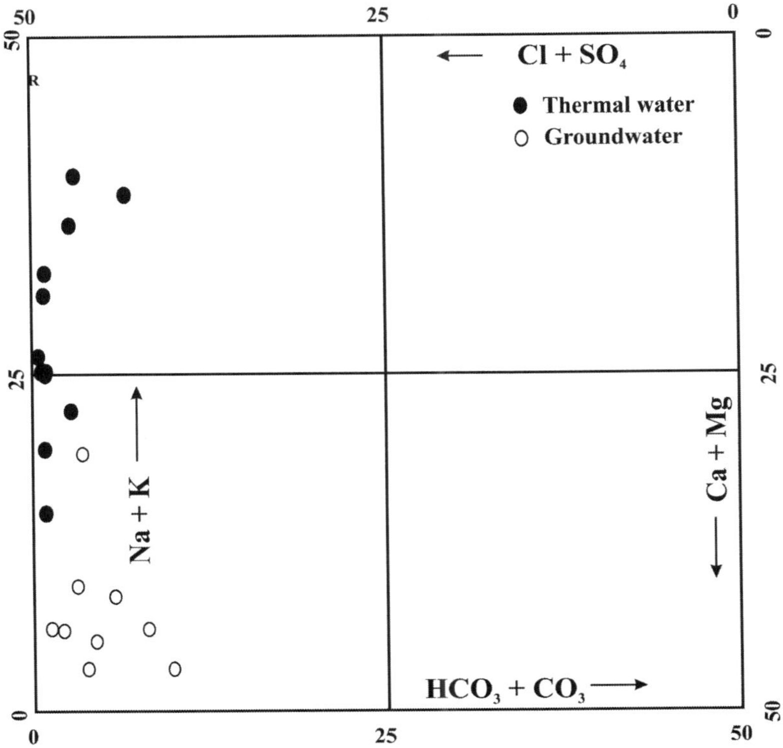

Figure 4.34 Langelier and Ludwig (1942) diagram showing the chemical characteristics of thermal, warm and cold waters from the Arabian Shield. Red Sea sample is also plotted for reference (adapted from Chandrasekharam et al., 2015c; Red Sea water from Banat et al., 2005).

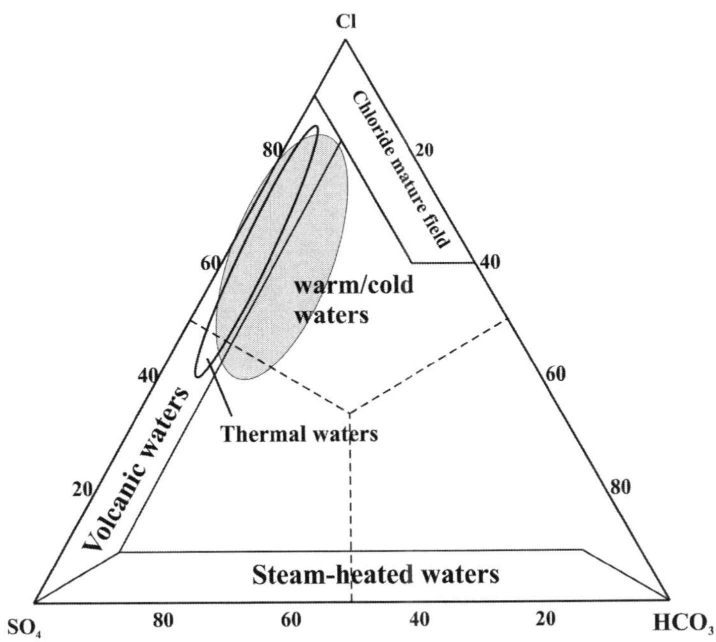

Figure 4.35 Cl-SO_4-HCO_3 diagram of Giggenbach (1988) showing the fields of thermal, warm and cold waters.

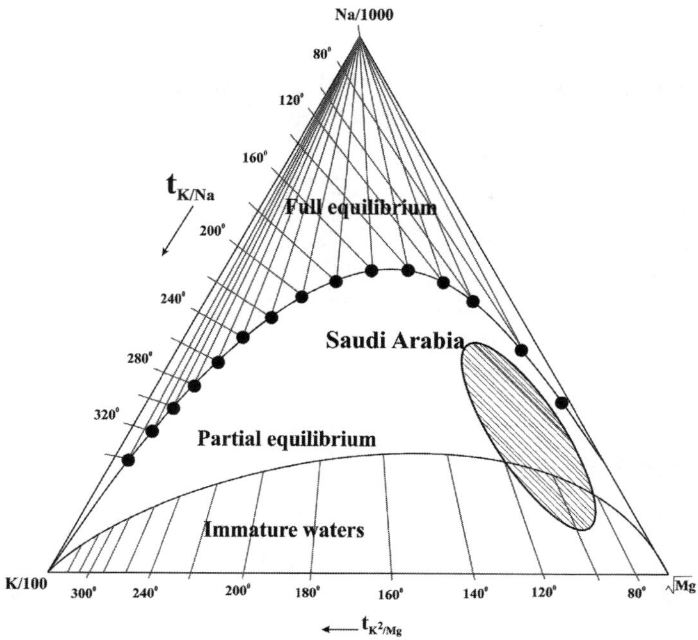

Figure 4.36 Na-K-Mg diagram (Giggenbach, 1988) showing the field of thermal waters from Saudi Arabian Shield.

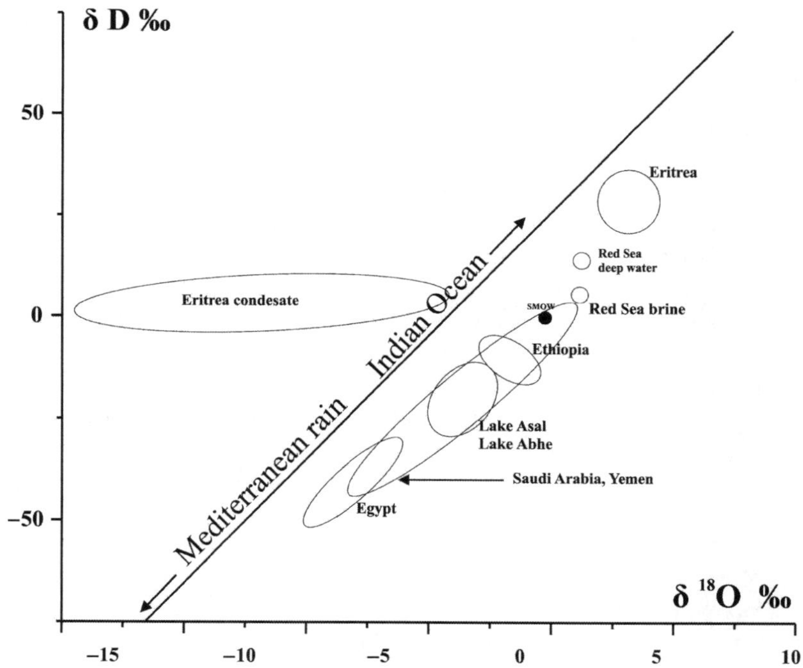

Figure 4.37 Oxygen and hydrogen isotope variation in thermal waters from geothermal provinces around the Red Sea (Mediterranean and Indian Ocean precipitation line is drawn from IAEA (2005)) (adapted from Chandrasekharam et al., 2015c).

to understand the behaviour of these isotopes in terms of rock water interaction and exchange of oxygen isotope between fluids and reservoir rocks. The presence of high temperature geothermal systems is supported by the distinct oxygen shift by the thermal waters from all the provinces. Except for Egypt's geothermal province, all others, as described above in the sections, are associated with or related to volcanic activity further supporting the presence of active high temperature geothermal systems in these regions. Oxygen exchange between silicates and fluids interacting with these minerals occurs at temperatures >220°C (Nuti, 1919). While meteoric water from the Indian Ocean's monsoon is the main feeder to the geothermal systems in Saudi Arabia, Ethiopia, Eritrea and Djibouti, a Mediterranean monsoon signature (IAEA, 2005) is imprinted on the geothermal systems of Egypt and partly Saudi Arabia. The position of the isotopic ratios in the steam condensate and thermal waters reported from Eritrea (Duffield et al., 1997, Lowenstern et al., 1999) appears to be from different sampling sites as these samples lie on different evolution lines. However, it is possible to make assumptions, based on steam condensate and thermal waters isotopic trajectories (Giggenbach, 1992, Nuti, 1999), that the steam condensate suggests a reservoir temperature of about 220°C. The oxygen shift shown by the thermal springs from all the geothermal provinces around the Red Sea suggests similar reservoir temperatures.

4.8 TECTONIC EVOLUTION OF THE LAND MASSES AROUND THE RED SEA

Egypt, Eritrea, Djibouti, the Republic of Yemen and Saudi Arabia were formed as separate land masses after the breakup of the Arabian-Nubian Shield at about 31 Ma due to the intense plume initiated volcanic and tectonic activity. The rifting was preceded by intense basaltic volcanism within a short period of 1.5 Ma (Hofmann et al., 1997, Coulie et al., 2003). The Red Sea represents a young basin floored by oceanic crust that developed due to the breakup of the Arabian and Nubian shields. The continental rifting and transition of ocean to continental crust dynamics are well preserved across the Red Sea axis between Eritrea and Saudi Arabia. The pre-rift configuration of the land masses and the associated rock types have already been discussed in the previous sections related to the respective countries. The plume was located below Afar and the plume head extended to regions below Eritrea, the Republic of Yemen and Saudi Arabia (Figure 4.38) (Bosworth et al., 2005, Camp and Roobol, 1992, Al-Amri, 1994). Initial magmatic activity that accompanied the plume activity was represented by intrusion of dike swarms along the entire length of the Arabian Shield. The geological structures that evolved over space and time that accompanied different stages of Red Sea evolution are preserved on either side over the Arabian Shield (including the Republic of Yemen) and Eritrea and Ethiopia (including Djibouti) (Figure 4.32). Based on the radiometric ages this activity occurred between 24–21 Ma (Feraud et al., 1991, Sebai et al., 1991). This was followed by intense basaltic volcanism over all the land masses (Egypt, Eritrea, Ethiopia, Djibouti, the Republic of Yemen and Saudi Arabia) between 21 and 14 Ma. The geothermal systems in all the above countries appear to have evolved during this active volcanic episode (Figure 4.39).

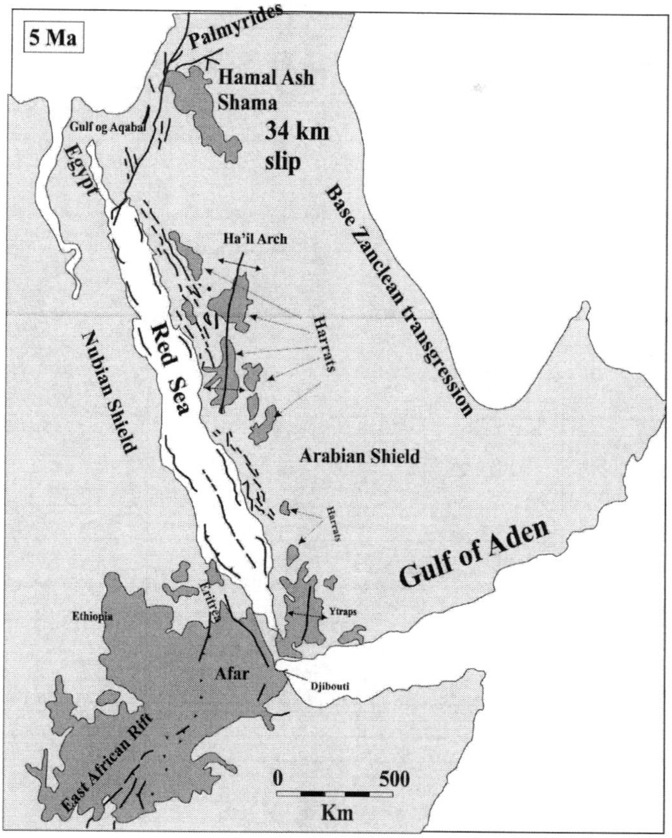

Figure 4.38 Position of Afar plume prior to the breakup of the Arabian-Nubian Shield at 31 Ma (adapted from Bosworth et al., 2005 and Chandrasekharam et al., 2015c).

The initial volcanic activity (plume related) was represented by Harrat as Sirat, H. Hadan, H. Uwaynd over the Arabian Shield and Yemen Traps in the Republic of Yemen while post rift volcanism is represented by H. Rahat, H. Lunayyir, H. Khaybar, H. Kishb, H. Nawassif and small eruptions in the Republic of Yemen (Figure 4.32). The volcanic activity over Djibouti and Eritrea started during the plume period. The volcanic activity is still continuing in all the countries around the Red Sea with the intensity of rifting of the axis increasing; the current spreading rate is reported to be 2 cm/y and this activity is still proceeding (Ueipass and Sylvie, 2012). The geothermal system associated with subduction zones and related volcanism is well documented and is well understood in terms of the heat source, composition of the geothermal fluids and gases, and the evolution of the geothermal systems. However, unlike such settings, the Red Sea geothermal systems evolved under a rift related magmatic setting and represent a set of unique geothermal provinces in the world.

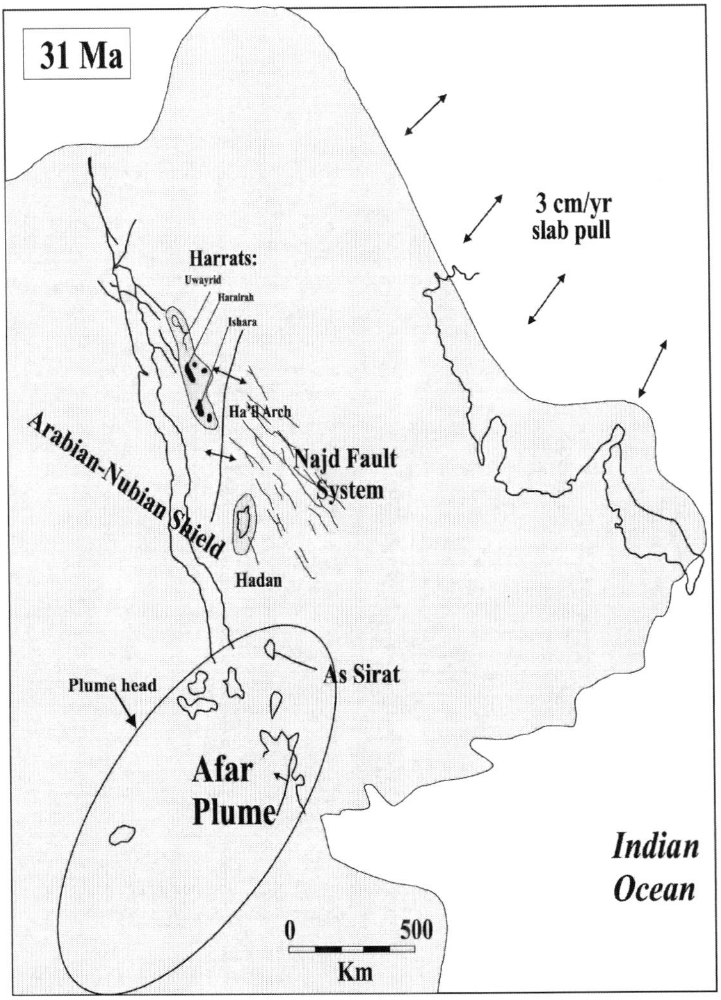

Figure 4.39 Co-evolution of basaltic volcanism over Egypt, Eritrea, Ethiopia, Djibouti and Saudi Arabia and geothermal systems (adapted from Bosworth et al., 2005 and Chandrasekharam et al., 2015c).

4.9 SUBSURFACE STRUCTURE ACROSS THE RED SEA AND CONTINENTAL MARGINS

A seismic refraction profile has been carried out across the Arabian Shield to understand the subsurface structure of the crust extending between the Red Sea axis and shield (Milkereit and Fluh, 1985, Debayle et al., 2001, Stern and Johnson, 2010, Girdler, 1970, Girdler and Evans, 1977, Gettings, 1982, Mooney et al., 1985, Pallister, 1987). The compressional wave velocity structure across the south western Arabian Shield, extending from Riyadh to the Farasan Bank in the Red Sea off the coast of

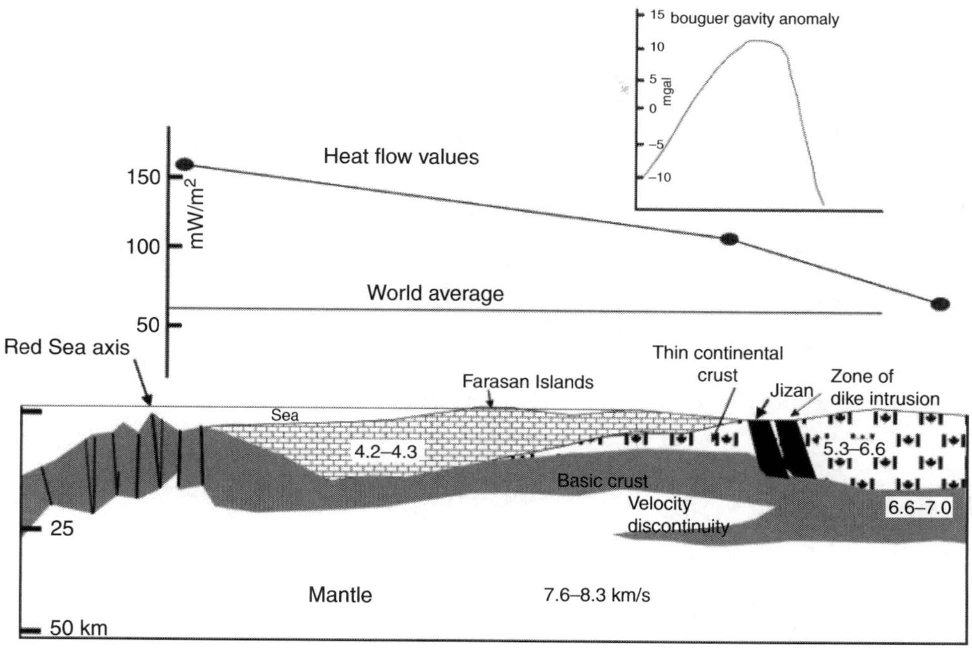

Figure 4.40 Schematic diagram showing the subsurface structures between the Red Sea axis and the shield axis (adapted from Chandrasekharam et al., 2015b).

Jizan, indicates the presence of two layers below the shield, each about 20 km thick with average velocities of 6.3 (top layer) and 7 km/s. The crust thins out towards the Red Sea margin from 40 km below the shield to 20 km. Above this layer lies the coastal plain and the ocean crust (Mooney et al., 1985). In addition to this, the surface wave tomography of fundamental and higher mode Rayleigh waves indicates the presence of a small low velocity anomaly deeply rooted in the upper mantle (Debayle et al., 2001). Most interesting findings based on ray tracing using synthetic seismograms recorded the presence of oceanic mantle material above the continental crust giving rise to double layer Moho (Milkereit and Fluh, 1985). Because of this strange structural setting, the region falling between the highlands over the shield and Red Sea axis recorded very high heat flow values indicating high geothermal gradients (Girdler, 1970, Gettings, 1982). Based on the available subsurface geophysical information, subsurface crustal cross-sections are drawn. A recent attenuation tomography model for the crust beneath Harrat Lunayyir recorded seismic swarm in 2009. The 3D attenuation model recognised a low attenuation zone down to 5 km depth corresponding to a rigid basalt cover. At greater depths, high attenuation anomalies were recorded that coincide with the seismicity cluster. This zone has been interpreted to be the conduit of an uprising magma. These magmatic fluids reached shallower levels (2 km) but were prevented from further rise and eruption by a rigid body (magma stoping) (Koulakov et al., 2014, 2015).

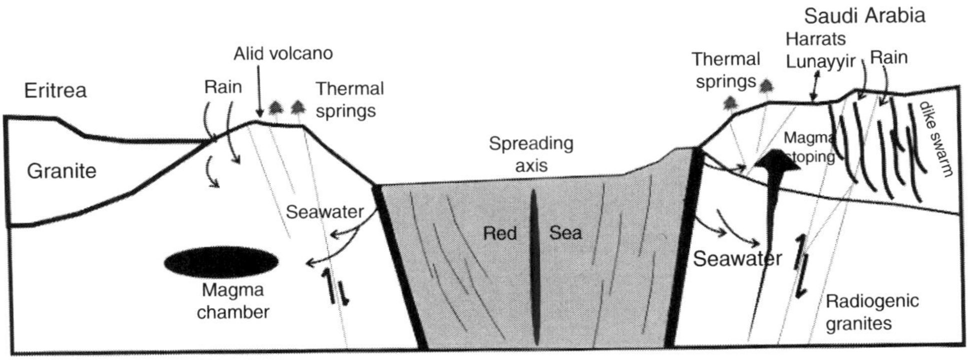

Figure 4.41 Schematic diagram showing the subsurface structures across Eritrea and the Saudi Arabian Shield (adapted from Chandrasekharam et al., 2015c).

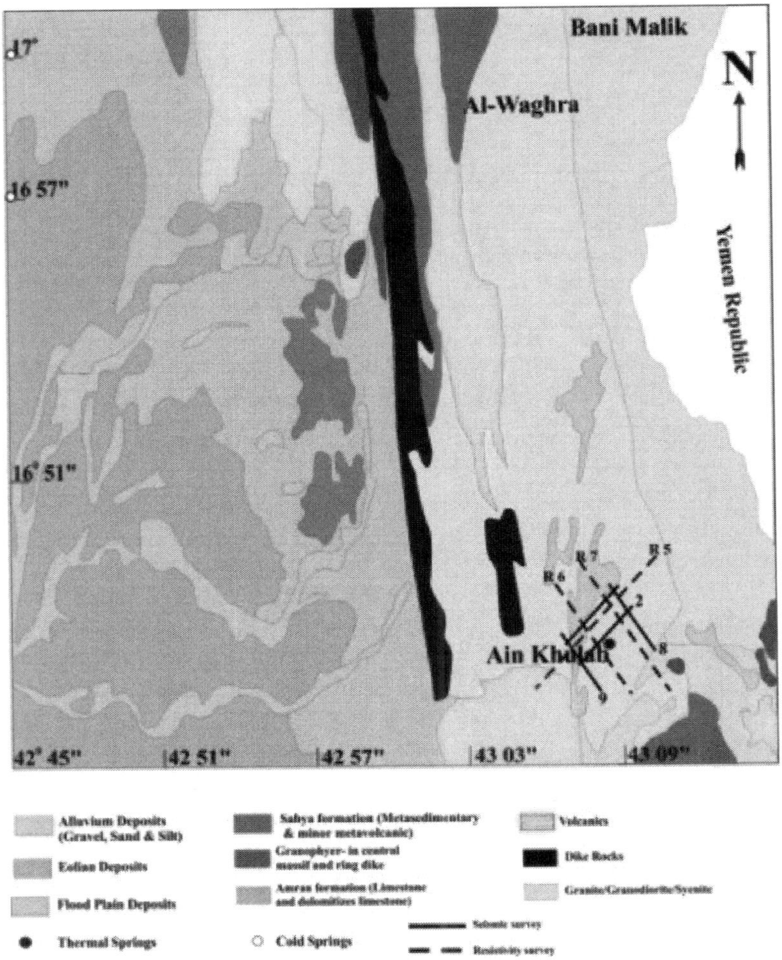

Figure 4.42 Geology and structure of Jizan geothermal province. R: Seismic reflection and R: Resistivity survey (adapted from Chandrasekharam et al., 2015b).

Bouguer gravity anomaly maps based on 2,200 gravity stations have been prepared along a 150 km wide and 8,850 km long strip over the Arabian Peninsula and the western shield extending from Riyadh to Farasan island in the Red Sea (Gettings, 1983). This gravity profile covers the seismic profile traverse of several workers (Milkereit and Fluh, 1985, Debayle et al., 2001, Stern and Johnson, 2010, Girdler, 1970, Girdler and Evans, 1977, Gettings, 1982, Mooney et al., 1985, Pallister, 1987). The gravity anomaly profile reveals several interesting subsurface features that are also reflected in the seismic profiles. The continental margin on the coastal plain is reflected by a steep gravity gradient. A model was developed combining the regional gravity anomaly pattern and seismic refraction profile discussed above. This model shows an upwelling of mantle material below the Red Sea and the presence of mantle at shallower levels below the Arabian plate. The presence of oceanic crust, sandwiched between the continental crust, was picked up by the seismic profiles and has been reflected in the gravity anomaly map (Gettings, 1983). A schematic cross section from the Red Sea axis to the shield and across the Red Sea extending between Eritrea and Saudi Arabian Shield has been prepared based on the seismic and gravity results and is shown in Figures 4.40 and 4.41.

Besides regional geophysical surveys, local seismic surveys have also been carried out to understand the local subsurface structures associated with the geothermal systems. Such an investigation was carried out over the Jizan geothermal province.

4.10 PATHWAYS OF GEOTHERMAL FLUIDS IN THE ARABIAN SHIELD

Geothermal systems in the SW part of the Arabian Shield are circulating through granites e.g. Al Lith and Jizan geothermal provinces (Figure 4.33) (Lashin et al., 2014,

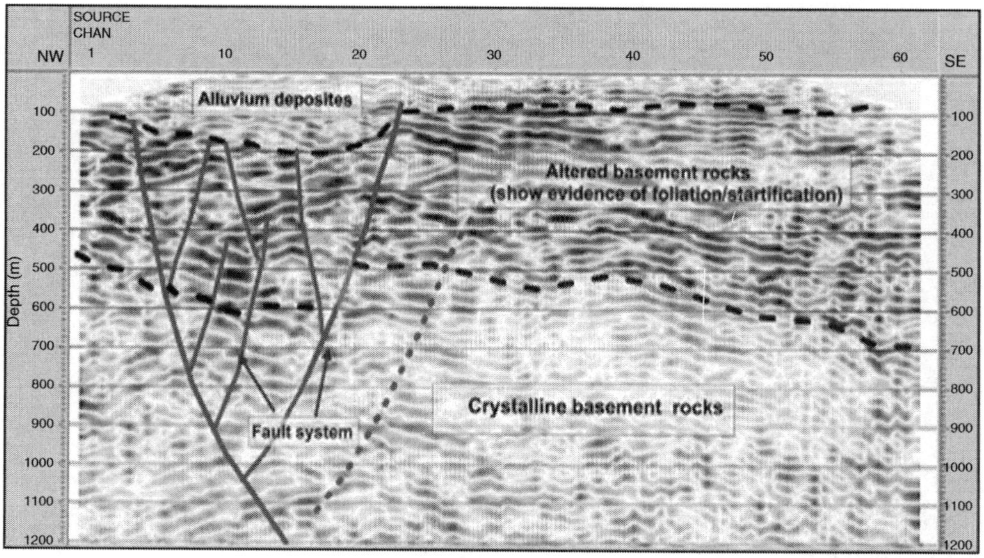

Figure 4.43 Seismic reflection profile along traverse 8 (NW-SE).

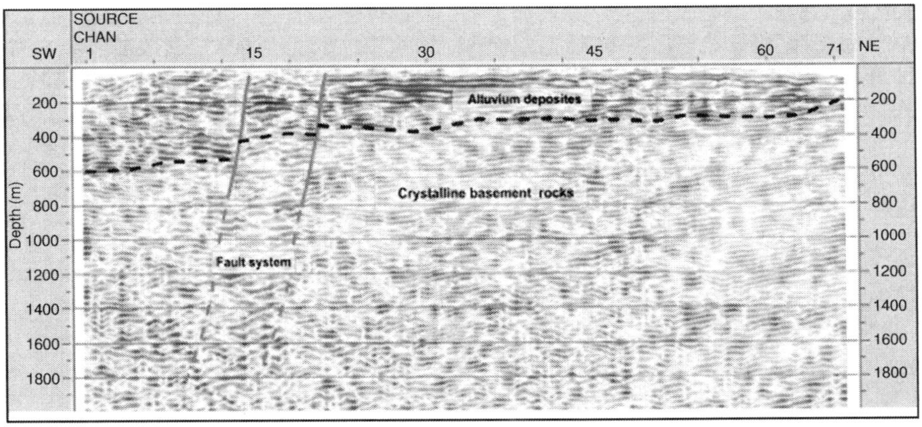

Figure 4.44 2D seismic reflection profile along traverse 2 (NE-SW).

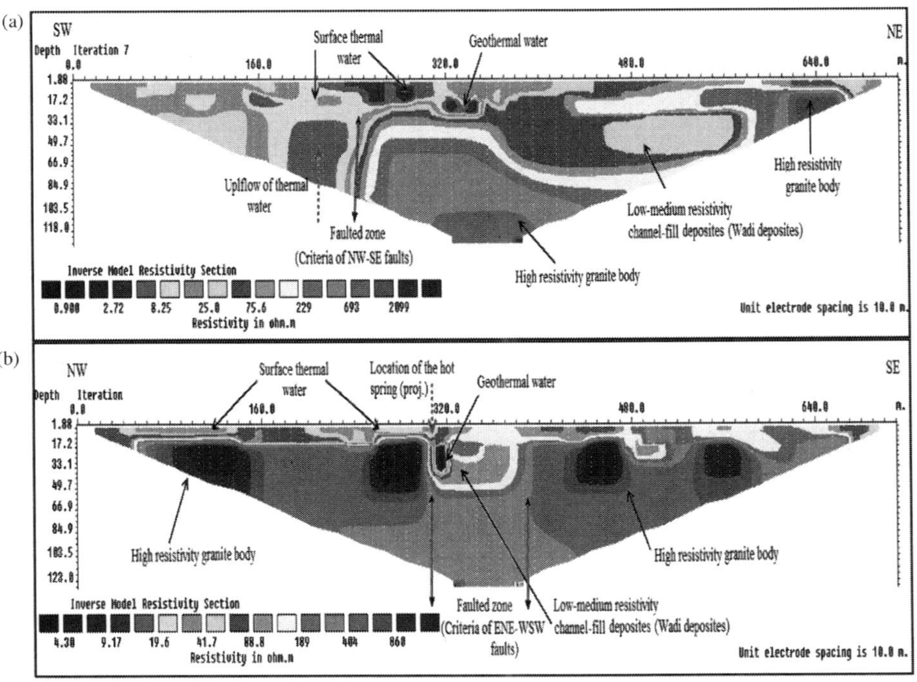

Figure 4.45 2D resistivity model showing the geothermal circulation system in Jizan province (adapted from Chandrasekharam et al., 2015b).

Chandrasekharam et al., 2015b). Seismic reflection and resistivity surveys have been carried out around Jizan geothermal province to understand the fluid circulation pattern and the geothermal reservoir characteristics in the granites. The geology and surface structural features around the geothermal province are shown in Figure 4.42.

The subsurface structures deduced across seismic traverses 8 and 2 (Figure 4.42) are shown in Figures 4.43 and 4.44.

The ENE-WSW transform faults with its characteristic flower structure and the NW-SE trending faults are clearly recorded in these two traverses (Figures 4.43 and 4.44). The 2D electrical resistivity model deduced from the electrical survey over the geothermal province is shown in Figure 4.45. The distribution and fault controlled circulation pattern of the geothermal fluids is clearly visible in Figure 4.45a.

The main ascending path of the thermal fluids is along the fractures located towards the SW side of the province. The topography of the granite batholith and the valley fill sediments are clearly distinguishable due to the contrasting resistivity values in Figure 4.45b.

Chapter 5

Carbon dioxide mitigation strategy

Focus on electricity decarbonisation. Strong policies supporting low carbon electricity could more than halve electricity emissions in 2030. This would save 9.5 Gt in 2030– an amount larger than China's total 2012 energy emissions –

(IEA, 2014d)

At the time of this book going to the press, a new United Nations climate agreement was going to negotiated by nearly 200 countries in Paris in December 2015. This agreement was to focus on reducing carbon emissions and limiting the increase in the global temperature to <2°C. The theme of the conference was the same as the IPCC Cancun meeting (COP18, 450 Scenario) held in 2012. In this meeting 195 countries attended the convention and only 38 countries agreed to control CO_2 (see Chapter 3, Figure 3.3). It is doubtful whether, within a short period of three years, another contention by a more or less similar number of countries that met in 2012, will bring in a drastic change in policy to control the increase in the global temperature (The Economist, June 1, 2015). Between the period from 1997, when the Kyoto Protocol was signed, and now, several policy actions have been proposed to contain CO_2 emissions, some of which have been adopted by several countries. The belief is that these actions will pay off in their own right and will make those 195 countries and other countries that did not participate in the Cancun meeting, to achieve post 2020 mitigation strategies more easily. The International Energy Agency (IEA) came out with five important actions that are expected to be a part of the COP21 agreement (IEA, 2014). These actions aim at the period between 2020 and 2030, which is crucial to bring a considerable change in the global CO_2 emission reductions and controlling the global temperature rise. These actions are 1) seize the benefit of immediate action to bend the global emissions curve, 2) focus on electricity decarbonisation, 3) reshape investment and accelerate innovation now in low carbon technologies, 4) mobilise non-climate goals to promote energy sector emission reductions and 5) strengthen energy sector resilience to climate change. The first action aims at bending the global emission curve (Figure 3.7) by a) adopting energy efficiency procedures, b) reducing electricity generation, c) reducing upstream methane gas emissions and d) doing away with fossil fuel subsidies. These sub actions will not affect a country's GDP and are hence easy to adopt. Energy efficiency will contribute a 49% reduction in CO_2 emissions. The action needed is to build energy efficient cooling and heating buildings. Here, geothermal energy can play a greater role, and countries can enforce policies to implement ground source heat pumps for this purpose. Germany and China are the leaders in implementing energy efficient geothermal systems. The installed capacity of geothermal heat use in Germany was 4,150 MWt in 2014 and 90% of this came from geothermal heat pumps (GHPs). The heat production in Germany in 2013 was about 5.5 tWh with 15% contributed by deep geothermal sources (Webse et al., 2015). GHPs can be installed anywhere in any country and it is therefore easy for all the countries to adopt this policy and contribute to the 49% reduction in CO_2 emissions. In addition to the building sectors, the industrial and transport sectors can also adopt this policy and help the global climate change.

A large volume of CO_2 emissions come from old inefficient coal based thermal power plants. These plants can either by upgraded or closed thereby saving about 21% of emissions. Here renewables play an important role since they can replace part of the electricity generation, which will no longer be generated due to the decommissioning of the old coal power plants. Under the renewables, geothermal energy can contribute a large part of the electricity since this source has the capacity to generate baseload unlike any other renewables and the geothermal power plants can operate 97% of the time with a life span of 25 to 30 years. Methane gas emissions from upstream oil fields can be contained by developing appropriate technologies. Energy subsidy on fossil fuel based sources should force countries to adopt any of the above sub- actions. In fact, like in Europe, energy subsidy should be given to renewable energy sources and governments should strictly implement this action without any hesitation. The Federal Government of Germany is offering new incentives to geothermal energy by increasing the subsidy to 0.25 €/kWh with an additional €0.05 for power generated from Enhanced Geothermal Systems (EGS) sources (Webse et al., 2015).

Under the action related to electricity decarbonisation, countries can, under the 2°C scenario, decrease emissions by 8 Gt by increasing the share of renewables in electricity generation and heat production. Since a large quantity of electricity is being utilised for generating fresh water from sea water in arid countries, renewables can play a greater role in meeting fresh water demand for the growing populations. Besides drinking water, fresh water use is expected to increase to support agriculture with the increase in population, especially in arid and Middle Eastern countries. Although coal cannot be replaced completely in the electricity sector, clean coal technology for electricity generation can be adopted and is being practised by several countries. Implementing carbon pricing (carbon taxes and carbon trading) will be a deterrent for many countries and will compel them to adapt decarbonised electricity generation technology. Although carbon pricing and trade are in place, a policy to strictly implement this strategy is lacking in many countries.

Evolving new technologies are being attempted by several countries, such as generating power from EGS and several countries are implementing these projects to generate electricity with zero carbon emissions. The last two actions will take shape once the above three actions are implemented by the countries with vigour. The outcome of the 2015 December COP21 meeting in Paris will provide clues as to whether the above actions are palatable to the majority of the countries or not. At least for those countries that have rich geothermal resources, implementing the above discussed actions to reduce CO_2 emissions should be easier, provided the respective governments in the such countries implement these actions.

By reducing carbon emissions, all the countries will, without doubt, benefit but the deterrent is the economic growth. Reducing carbon emission affects the socio-economic growth of several countries, effectively the rich countries will become richer. The proposed solution to this problem is to impose uniform carbon tax on countries emitting large quantities of CO_2 as was discussed in the 2009 Copenhagen UN climate change convention. But this remained as a proposal although certain governments imposed carbon tax. These pledges and reviews discussed year after year by the countries are not a permanent solution to achieve controlling the 2°C rise in the global temperature by 2020. In the absence of a uniform carbon tax and binding commitments by countries, such pledges will continue in the future (The Economist, 2015).

In the case of all the countries located around the Red Sea, as discussed in Chapter 4, the respective governments should adopt and implement these policies to explore and utilise the geothermal energy resources and reduce dependency on fossil fuels. For example, countries like Djibouti, Eritrea and Saudi Arabia depend entirely on fossil fuels to support their energy demand. In addition, Saudi Arabia utilises a considerable quantity of fossil fuel to generate fresh water through desalination processes (see Chapters 3 and 4).

Eritrea and Djibouti, with their low population and low CO_2 emissions, can become zero carbon countries by developing their geothermal energy resources under the action proposed by IEA (2014) that is anticipated to be adopted by the COP21 in Paris. Similarly, Saudi Arabia can immediately save 13 Mt of CO_2 by switching over to geothermal energy supported desalination processes along the entire length of the eastern margin of the Red Sea and can become a fresh water surplus country amongst the Middle-East countries. The subsidy on fresh water can be removed once geothermal energy supported desalination plants operate in the country. Apart from supplying drinking water to a population of 29 million, Saudi Arabia can support the development of its agricultural sector to enhance its barley and wheat production (see Chapter 3) and provide food security to its growing population.

Chapter 6

Geothermal exploration techniques

The shortage of energy is a myth that persists today. Clean, renewable energy is abundant on our planet –
yet marginally utilized. The technology exists today to tap a much greater share of "green energy" –
and reduce atmospheric pollution and the greenhouse gases from burning fossil fuels.
Peter Meisen, President, Global Energy Network Institute (GENI), 1997

6.1 GEOTHERMAL SYSTEMS

Geothermal systems can broadly be grouped into two types: 1) hydrothermal systems and 2) enhanced geothermal systems (EGS). In the first group, meteoric water is the main source feeding the circulatory system in the majority of the cases, while in EGS water or CO_2 is circulated to extract the heat from the hot rocks that are devoid of any circulating fluids. In the case of hydrothermal systems, the heat source is either a magma body below the surface or deep seated faults that permit deep circulation of the fluids that get heated due to prevailing geothermal gradients or sometimes high geothermal gradients. The meteoric waters, that percolate deeper into the crust channelled by fractures and faults, get heated and rise to the surface, due to buoyancy through the fractures and faults, and emerges as hot springs. In the case of EGS, the rocks, usually of granitic composition, generate high heat due to the presence of radioactive elements like uranium (U), thorium (Th) and potassium (K). This heat can be extracted by inducing a system of fractures in these rocks and circulating water through this interconnected fracture system. The temperatures of these rocks vary between 100 to 200°C depending on the quantity of radioactive elements present in these rocks. Generally, at depths of about 3 to 5 km, these rocks attain such temperatures and hence, while the hydrothermal systems are site specific, the EGS are present anywhere on the earth. With the commissioning of a pilot power plant based on EGS technology in Soultz (France), EGS technology is maturing and appears to be the most promising energy source with a zero carbon foot print (MIT, 2006). Now, instead of water, carbon dioxide is being circulated through these induced fractures in granites to extract heat. This technique thus serves two purposes: generating carbon-free energy and sequestering carbon dioxide that is generated from fossil fuels based power plants. In this section hydrothermal systems will be discussed followed by EGS.

6.1.1 Characteristics of geothermal fields

Understanding the surface geological and tectonic regime of any geothermal province is vital in any geothermal exploration programme. These provinces are developed over a variety of unique geological and tectonic settings; their evolution is interrelated. In conjunction with the surface geology and tectonics, the subsurface characteristics of the provinces provide further clues to the circulating pattern of the fluids and their interaction with the subsurface geology and structure. This information may not be readily available, hence indirect methods like seismic, magneto telluric and resistivity surveys play an important role in deciphering such information. Such investigations help in identifying the right location for further exploration through exploratory drilling thereby minimising the number of exploratory bore holes and the drilling cost. As listed below, the hydrothermal systems can be classified broadly into six groups:

1 subduction-related volcanic settings – ocean *vs.* ocean plate involvement, ocean *vs.* continental plate involvement
2 continental collision zones
3 infra-continental rifts and ocean rifts related volcanic settings
4 infra-continental rifts without volcanism
5 geo-pressured systems associated with oil fields
6 crust-mantle boundary disequilibrium settings

The characteristics of the systems associated with the above groups are briefly discussed with examples from around the world. This information is essential to understand the evolution of geothermal systems under a given set of geological and tectonic conditions.

6.1.1.1 Subduction-related volcanic settings

Arc volcanism is common under ocean vs. ocean plate subduction and ocean vs. continental plate subduction. Both these settings are loci of geothermal manifestations, generally of high temperatures, associated with emission of thermal water, steam and fumaroles. The geothermal systems in Indonesia and the Philippines are good examples of these settings. In Indonesia, Sumatra alone hosts nearly 200 volcanoes; many of them are active. The geothermal system here is represented by steaming pools and steaming ground. The tectonic setting and associated geothermal fields are shown in Figure 6.1.

Detailed lithological characteristics of the islands have been reported by several authors: Stauffer, 1983, Cooper et al., 1989, Page et al., 1979, Mitchell, 1993. The Indo-Australian Plate converges below the Java Trench at about 60 mm/year (Minster and Jordan, 1978). The Sumatra Fault and the geothermal fields lie parallel to the axis of the subduction line. The geothermal fields are associated with active volcanic zones in all the islands. Indonesia has an estimated 27,000 MW of geothermal potential and only 4% of this is currently being exploited (Riki et al., 2005).

6.1.1.2 Continental collision zones

The best example of geothermal systems associated with continent to continent collision zone is the Himalayan geothermal province (HGP). The entire collision zone,

Figure 6.1 Steaming pool and ground around Sorikmarapi volcano, Sumatra, Indonesia (Photo by D. Chandrasekharam).

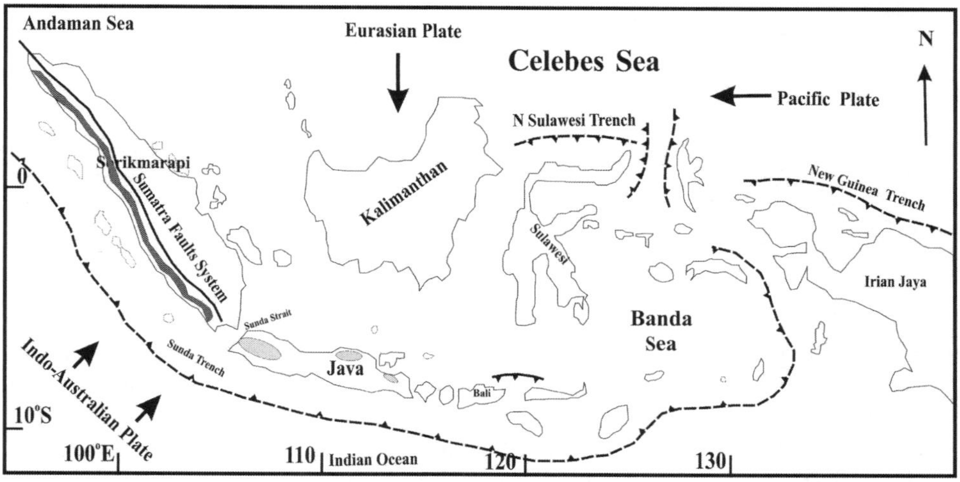

Figure 6.2 Tectonic map of Indonesia showing major geothermal fields (adapted from Chandrasekharam and Bundschuh, 2008).

extending from NW to E Himalayas, is the locus for geothermal manifestations, with the presence of more than 100 thermal springs and steam vents. The length of the HGP is about 1,500 km (Chandrasekharam and Bundschuh, 2008). The presence of granitic melts at shallower levels and shallow mantle depth and the frictional heat resulting from thrust are the main source of heat for geothermal systems. Snow melt is the

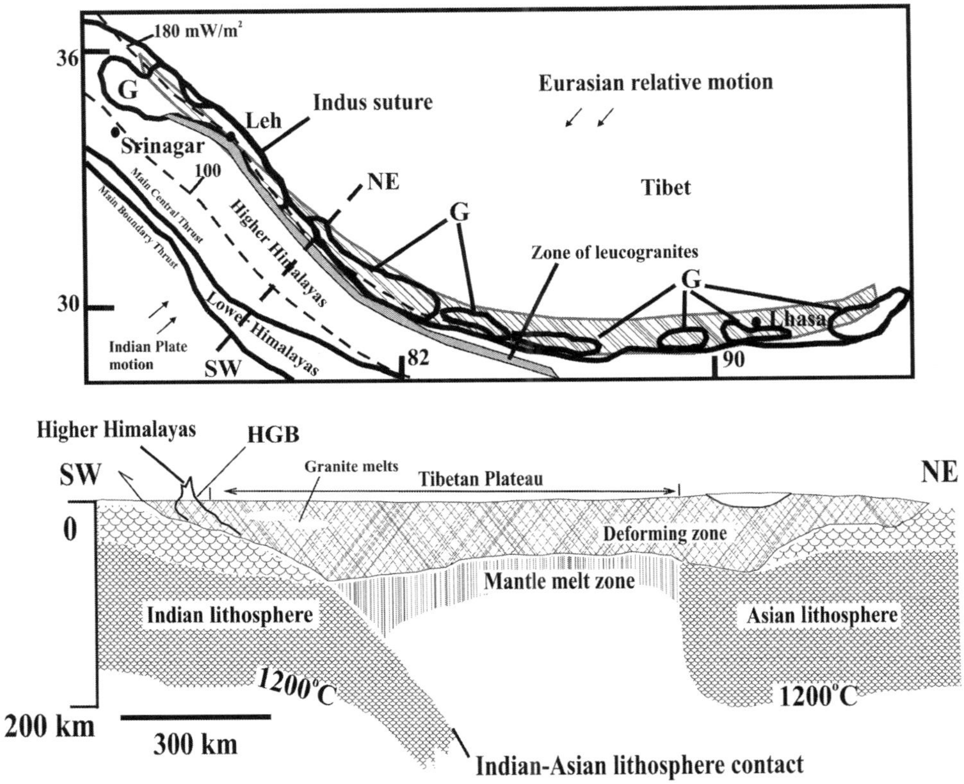

Figure 6.3 The cross section shows the position of the crust, shallow anatectic granite melts. The cross section is drawn across the line SW-NE. The leucogranites are the main host rock for the HGP (adopted from Chandrasekharam and Bundschuh, 2008).

main recharge and the thermal water circulation is within the granites. Low $^3He/^4He$ ratios indicate granites as the main reservoir in this province (Chandrasekharam and Bundschuh, 2008). The granites exposed along the Indus suture zone are younger and their ages range from 53 to 60 Ma (Schneider et al. 1999a, b, c, Searle 1999a, b, Le Fort and Rai 1999, Harrison et al., 1998, 1999). Due to regional thrust exerted by the Indian Plate (Figure 6.3), these granites have been highly fractured and are thus highly permeable. The fluids circulating through the fractures scavenge the heat from these granites. In fact, the HGP is controlled by the granites and thus represents a natural EGS system in operation.

Granites melts, generated due to anatexis at shallow crustal levels, the presence of radioactive elements in the younger granites and shear tectonics are the main driving forces for the geothermal systems in HGP. The Yangbajing geothermal field in Tibet, located on the HGP, is generating about 28 MWe (Dor Ji and Ping, 2000); the hot aquifer is located in the granites at different depths with increasing reservoir temperatures (Figure 6.4).

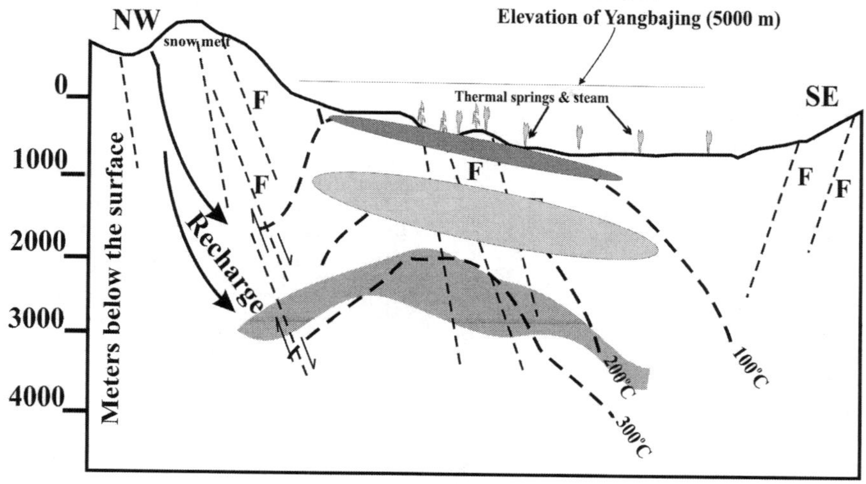

Figure 6.4 Schematic cross section showing the thermal fluid circulation system and thermal reservoirs in Yangbajing. The presence of three reservoirs has been established based on geophysical investigations (adapted from Dor Ji and Ping, 2000).

6.1.1.3 Infra-continental rifts and ocean rift related volcanic settings

The best examples of this type of geothermal province are located in Ethiopia, Kenya, Djibouti, Eritrea, Egypt and Saudi Arabia. The geological and tectonic settings of Ethiopia and Kenya are discussed below and the geothermal provinces from other countries have already been discussed in earlier chapters. In fact, the East African Rift, which originates from Afar and extends into Ethiopia, extends further south into Kenya and forms an important domain for high temperature geothermal systems. Besides the Tendaho geothermal province, discussed in Chapter 4, the Ethiopian rift (part of the East African Rift system) hosts several other geothermal systems located at Aluto Langano, Corbetti, Abaya, Tulu Moye, Fantale, Kone and Dofan (Figure 4.17). Among these fields, Aluto Langano and Tendaho are the most promising high temperatures fields. Aluto Langano is located towards the south in the rift valley. Exploratory drilling data indicate reservoir temperatures of 180°C at 1,800 m and 335 at 2,200 m. This field has the potential to generate 100 MWe but now the initial 8.4 MWe plant is being expanded to generate 70 MWe (Tassew, 2015).

The East African Rift, which starts from the Afar, extends into Kenya. While geothermal development is slow in Ethiopia, Kenya has forged ahead in establishing a sizable quantity of geothermal power and is expanding aggressively to exploit its potential of 10,000 MWe (Omenda and Simiyu, 2015). The Kenya Rift Valley hosts about 11 geothermal sites and the most prominent ones are shown in Figure 6.5.

The most exploited field is the Olkaria. The Olkaria geothermal field is located in the Olkaria volcanic complex dominated by pyroclastics and rhyolites that were erupted 180 years ago (Figure 6.6). This field has the capacity to generate 1,500 MWe. From an initial 15 MWe, the generation has increased to 140 MWe and in addition

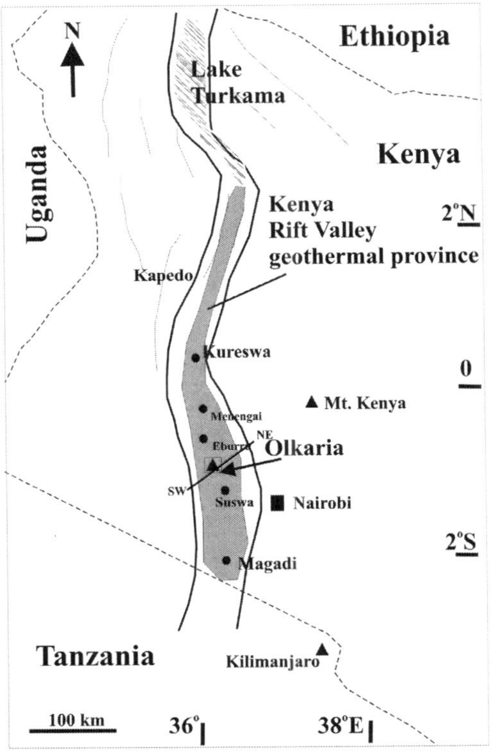

Figure 6.5 Geothermal provinces of Kenya (adapted from Omenda and Simiyu, 2015).

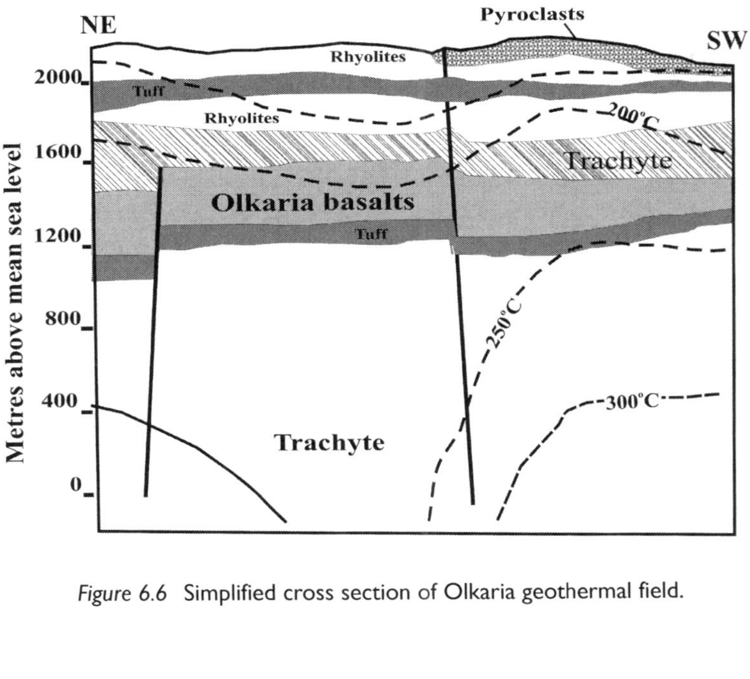

Figure 6.6 Simplified cross section of Olkaria geothermal field.

70 MWe well head generators have been installed. Steam is captured from wells drilled to depths between 900 to 2,200 m (Omenda and Simiyu, 2015). The geothermal power plants operate inside the wild life sanctuary and thus demonstrate the environmentally-friendly nature of this green energy source.

The geothermal provinces associated with an ocean rift related volcanic tectonic setting are those occurring around the Red Sea. The geothermal provinces in Eritrea, Djibouti, Ethiopia, Yemen, Egypt and Saudi Arabia all fall under this category. The geological and tectonic settings controlling the geothermal fluid circulation, their chemical evolution and resource potential have been discussed in detail in Chapter 4.

6.1.1.4 Infra-continental rifts

Precambrian shields and ancient crystalline platforms that are not associated with active volcanic manifestations also host geothermal systems. They are represented by those occurring in Larderello, Italy and in Cambay, Jharkhand, Tattapani and west coast in India (Figure 6.8).

The geothermal field at Larderello produces superheated steam due to a mass transfer of heat from the rocks to the fluid (Minissale, 1991). This is one of the few geothermal systems in the world that produces superheated steam. The quantity of steam produced in Larderello is of the order of 26 Mt/year (million tonnes) (Chierici, 1964) while the power plant started working since the last 100 years. At present Italy is generating 914 MWe out of which 100 MWe is generated from the Larderello field (Razzano and Cei, 2015).

The steam is stored in the reservoir as water (Marconcini et al. 1962, James, 1968). Tectonically, this area is characterised by extensional tectonics since the Late Miocene, the period that coincides with the late stage opening of the Red Sea and the initiation of the sea floor spreading along the Red Sea axis. During the extensional regime, the area experienced a sporadic compressional regime during Pliocene. Such tectonic configuration has resulted in block faulting and with the fault axes trending NW-SE and along the axis of these structure granitic magma was emplaced. The anomalous geothermal gradient and high heat flow in the Larderello field is related to the emplacement of these granitic batholiths (Minissale, 1991). A schematic cross section showing the subsurface tectonic configuration of the Larderello field is shown in Figure 6.7.

Another province that belongs to this group is the Son-Narmada-Tapi Lineament (SONATA) in India (Figure 6.8). The SONATA is represented by the Narmada-Tapi mid continental rift system which extends from Gujarat in the west to west Bengal in the east. It was formed due to the interaction between two proto-continents (Naqvi et al., 1974, 1978) during the early stages of the Indian Plate development and reactivated after the collision of India with the Eurasian Plate which triggered the Himalayan Orogeny (Figure 6.8).

The most prominent high temperature geothermal fields over SONATA are located in Tattapani, Bakreswar and Tantloi on the eastern side of SONATA and the Cambay geothermal province at the western end of the rift. The deep sounding seismic (DSS) profile across the SONATA (Kaila et al., 1981), south of Tattapani, indicates that the faults extend up to the mantle depths although the thermal gases and thermal waters of Tattapani that flow through these faults recorded very high helium content (0.5 to 7%) with a very low $^{3}He/^{4}He$ ratio (Minissale et al., 2000). The surface temperature

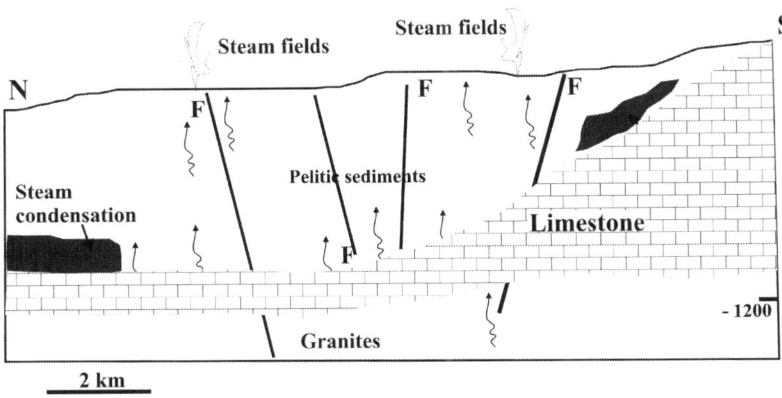

Figure 6.7 Subsurface structure of Larderello geothermal field (adapted from after Minissale, 1991).

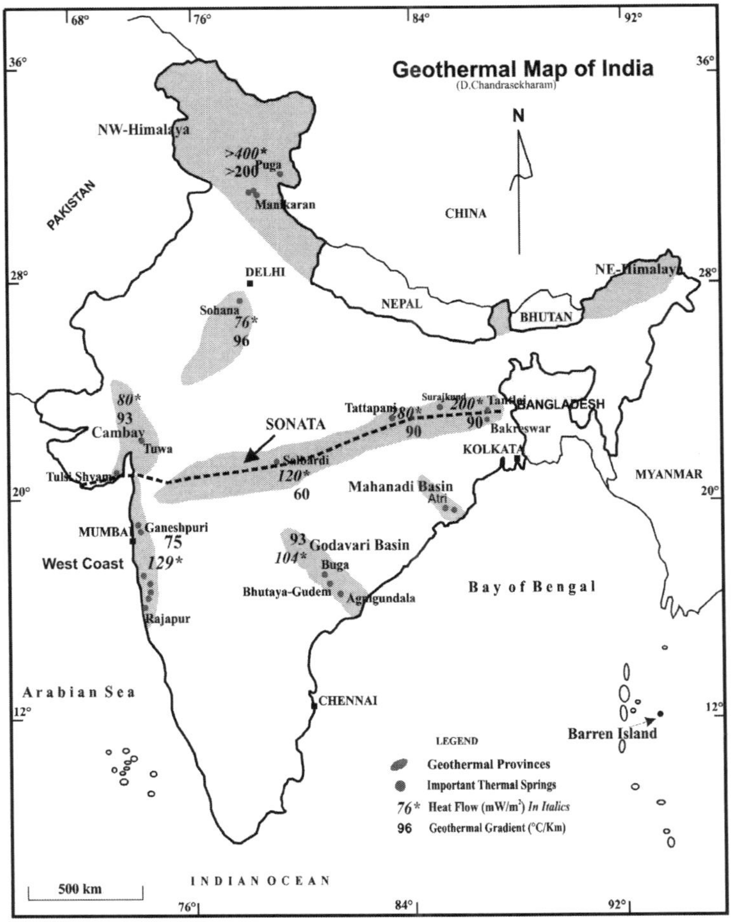

Figure 6.8 Geothermal provinces of India showing the SONATA rift related geothermal fields (adapted from Chandrasekharam and Bundschuh, 2008).

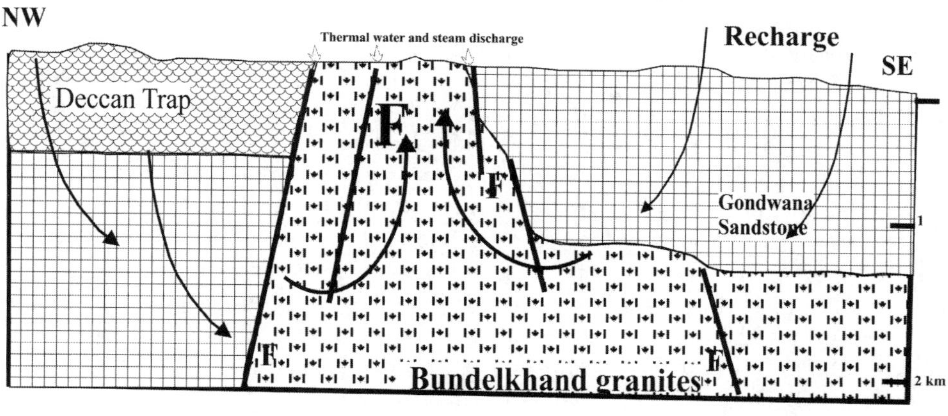

Figure 6.9 Schematic cross section showing the circulation of thermal fluids in the Tattapani thermal field. The Bundelkhand granites contain high radioactive elements and generate >8 μW/m^3 of heat (adapted from Minissale et al., 2000, Chandrasekharam and Antu, 1995, Chandrasekharam and Chandrasekhar, 2010b).

of the thermal waters is 93°C and the flow rate of the springs is 50 L/s. The entire SONATA has a sedimentary insulation over a high heat generating granite that lies at about 2 km depth, giving a geothermal gradient of 90°C/km (Chandrasekharam and Antu, 1995, Chandrasekharam and Prasad, 1998). A schematic cross section of the Tattapani geothermal field is shown in Figure 6.9.

The Bakreswar and Tantloi geothermal provinces, located on the eastern edge of SONATA (Fig. 6.8), host several thermal springs flowing through granites with surface temperatures varying from 86 to 90°C. These springs, and associated steam, are of high helium content; the helium is being extracted on a commercial basis from the thermal discharges (Chaudhuri et al., 2015). A thermal reservoir is located within the granites and a gravity spectral analysis indicates structural discontinuity at about 600 m either due to mantle up warp (as a consequence of the Rajmahal volcanic eruption) or basic dike intrusion. The estimated reservoir temperatures vary from 132 to 250°C and the estimated depth of the thermal reservoir, based on silica thermometry and a geothermal gradient of 90°C/km, is located at about 1.5 km (Singh et al., 2015).

The geothermal provinces that fall under the last two groups (geo-pressured systems associated with oil fields and crust-mantle boundary disequilibrium settings: see section 6.1.1) are not elaborated here. The geo-pressure geothermal systems are very common in oil and gas fields with the thermal waters attaining temperatures above 100°C and sometimes associated with steam emanations. Since these fields are controlled by the several oil companies, sufficient data is not available to make any reasonable inferences on such thermal systems. Morphometric analysis indicates an upwelling of the asthenosphere below Mongolia, creating a thermal anomaly and giving rise to a high heat flow. The temperatures of the thermal springs emerging to the surface vary between 20 to 90°C and thermal waters are being used for district heating systems (Chandrasekharam and Bundschuh, 2008).

6.2 GEOCHEMICAL EXPLORATION METHODS

Before commencing an exploratory drilling programme in any geothermal project, information related to the evolution of the thermal waters in terms of the geological and tectonic framework associated with the geothermal systems, the chemistry of the thermal fluids and thermal gases, the circulation pattern of the fluids within the reservoir rocks, flow rates, the aquifer parameters as reflected by the chemical constituents in the thermal waters, and subsurface geological and tectonic features deduced from geophysical survey are all very important. Such information is important for the evaluation of any geothermal site, and with certain sound assumptions it is possible to estimate the power that can be generated from a site. These techniques have been explained in detail by several researchers in case studies: Ellis and Mahon, 1977, Truesdell, 1976, Truesdell and Hulston, 1980, Fournier and Rowe, 1966, Tonani, 1980, Giggenbach et al., 1983. This information, even though it is preliminary and may not be useful for drawing firm conclusions, is essential for working out the preliminary project costs. Such investigations reduce the number of exploratory drill holes required for estimation of the project cost. Warner Giggenbach (1988) developed a reasonable geochemical exploration template for geothermal systems to quickly judge the characteristics and potential of a thermal field. These techniques were applied for the geothermal sites around the Red Sea and discussed in Chapter 4; the principles behind these techniques are deliberated briefly here.

6.2.1 Classification of geothermal waters

The chemical composition of geothermal waters is controlled by the rocks with which the fluid is in contact. The main source that feeds geothermal systems is rain water, which percolates into deeper crustal levels and gets heated due to the prevailing geothermal gradient. Thus, the thermal waters are basically groundwater occurring at deeper levels which have been circulating within the rocks for thousands of years. Once heated, the water tends to come to the surface, due to buoyancy, through interconnected fracture systems in the rocks that act as channels. These waters may ascend to the surface fast or slowly from the reservoir, depending on the pathways. Fast ascending waters, in general, retain their original chemical signature, while waters ascending slowly have time to interact with a variety of rock types that they encounter en-route and thus undergo a change in composition. During this time, they may precipitate certain dissolved constituents or dissolve certain new constituents. Further, at the near surface environment, these ascending waters may encounter the groundwater table and mix with these young cold groundwaters before emerging to the surface. Those thermal waters that attain high temperatures in the reservoir, or at deeper levels, while ascending to a low pressure environment near the surface, may boil, due to a separation of vapour. The separation of vapour phase also influences the chemistry of the thermal waters. In case the circulation is taking place within an active volcanic system, then chemical components from the magma will also find their way into the circulating thermal fluids. Thus, a variety of processes operate throughout the circulation of the thermal waters from the time they enter the earth's crust. A schematic diagram showing a typical geothermal system is shown in Figure 6.10.

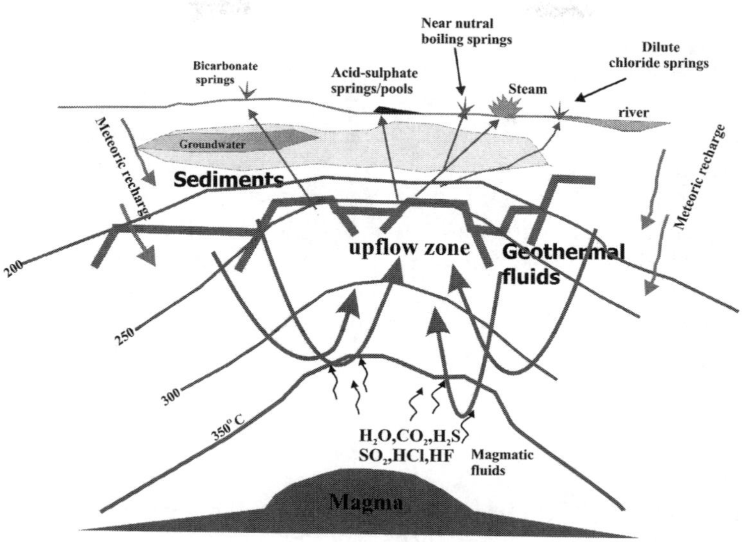

Figure 6.10 Schematic diagram showing a typical geothermal system.

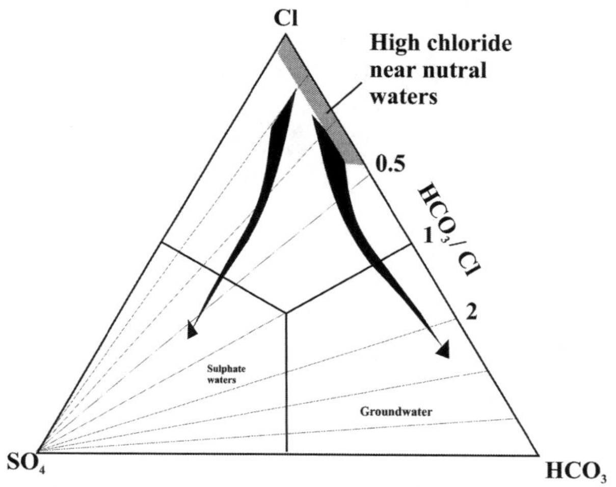

Figure 6.11 Giggenbach's Cl-SO_4-HCO_3 diagram showing the classification of thermal waters (adapted from Giggenbach, 1988).

Based on a large number of samples from thermal springs and thermal waters from bore wells, Giggenbach (1988) proposed an anion diagram that gives an idea about the waters that attained near equilibrium with the reservoir rocks and those that undergo chemical change due to their interaction with the rocks or groundwater before emerging. This diagram is shown in Figure 6.11.

6.2.1.1 Anions

After surface sampling, those thermal water samples that fall in the shaded region in Figure 6.11 are considered for estimating the reservoir temperature. Those samples that shift towards the HCO_3 field indicate their interaction with the groundwater before emerging to the surface. In certain geothermal fields, the thermal waters, depending on the subsurface lithological configuration, may flow along the boundary of the shallow aquifer there by heating the aquifer and the groundwater. Such waters also emerge to the surface as thermal waters and are, in a large number of cases, mistaken for original thermal waters. These thermally-heated groundwaters however, fall in the groundwater field in Figure 6.11. In active volcanic sites, the SO_2 from the volcanic gases mixes with the circulating thermal waters thus enriching them with SO_4. These samples shift towards the SO_4 field in Figure 6.11. This diagram discriminates original thermal waters from mixed waters.

6.2.1.2 Silica

Solubility of silica minerals in thermal waters is linear with temperature and combined with pressure the solubility increases further by about 19% (Fournier and Potter, 1982). The understanding of the solubility of silica in thermal waters is important because silica are often used as a chemical thermometer to estimate a reservoir's temperature. In the case of hot fluids with temperatures of about 250°C, ascending to the surface fast, dissolved silica do not precipitate significantly. Hence dissolved silica in such waters assumes importance in estimating the reservoir temperatures. However, as discussed above, the silica solubility varies with temperature, pressure and also the salinity of the fluids. The following equation is often used to calculate the solubility of silica in saline thermal waters between a 25 and 900°C temperature range (Fournier, 1983).

$$\log m = A + B(-\log \rho F) + C(-\log \rho F)^2$$

where
$A = -4.66206 + 0.0034063T + 2179.7T^{-3} - 1.1292 \times 106T^{-2} + 1.3543 \times 108T^{-3}$
$B = -0.0014180T - 806.9T^{-1}, C = 3.9465 \times 10^{-4}T$
ρ = density of the solution, F = weight fraction of water in the solution
m = molality of dissolved silica, T = temperature in Kelvin

In the case of slow ascending thermal waters, steam separates, often at the near surface environment due to adiabatic cooling. Silica content tends to increase in the fluid phase due to adiabatic cooling and hence needs some correction for estimating the reservoir temperatures (Figure 6.12). A similar concentration change in silica occurs when ascending thermal waters mix with cold groundwaters. In such cases, two situations can be visualised. In the first case, mixing can take place after adiabatic cooling or adiabatic cooling may set in after mixing or simultaneously. In such cases silica content can be estimated using Figures 6.13a and b.

The silica content corrected for steam loss may be used for estimating the reservoir temperatures using silica thermometers. In most of the geothermal fields (associated with volcanic activity), CO_2 is a component in the steam. When steam separates from the system, the thermal water pH increases thereby affecting the solubility of silica. Along with steam correction, pH correction should also be carried out in order to

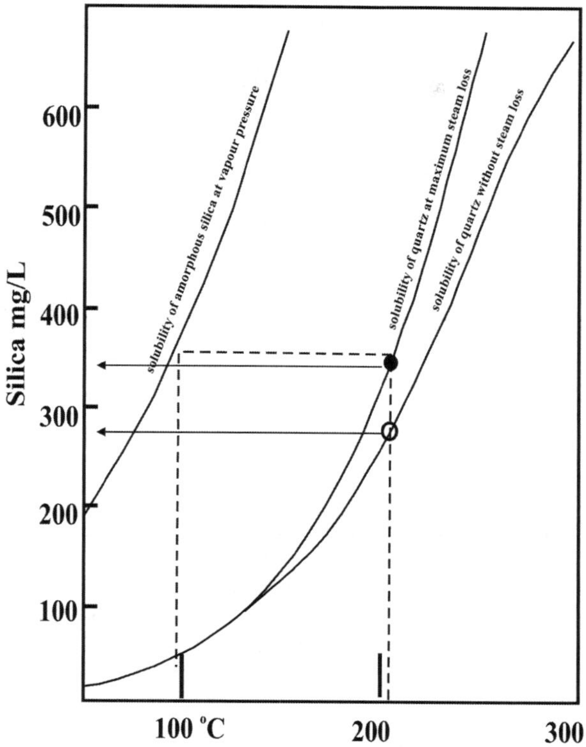

Figure 6.12 Estimation of silica content in thermal waters due to adiabatic cooling. Open circle: concentration of silica at 210°C before steam separation. Solid circle: concentration of silica after steam separation at near surface environment (adapted from Chandrasekharam and Bundschuh, 2008).

obtain the correct silica values (Fournier, 1985). Once the corrections for silica content are applied, reservoir temperatures can be calculated using established temperature related solubility of silica equations (Fournier and Rowe, 1966, Fournier, 1991).

6.2.1.3 Cations

While silica content in thermal waters is a function of temperature, the cations content is related to the exchange reactions and redox conditions prevailing during the reaction. The minerals that are involved are Na and K feldspars and micas (including pyroxenes and amphiboles which contribute Mg). The concentration of cations reflects the reaction that have taken place in the reservoirs between the rocks and fluids at high temperatures; it is therefore widely used for estimating the reservoir temperatures (Ellis and Mahon, 1967, Fournier and Truesdell, 1973, Giggenbach, 1988). These two thermometers, Na-K and K-Mg, generally give different reservoir temperature estimates because of different rates at which the reactions involving Na-K and Mg (felspar and micas) re-equilibrate under the new physico-chemical environment during the ascent of the fluids to the surface. Reactions involving K and Mg are sensitive to conditions

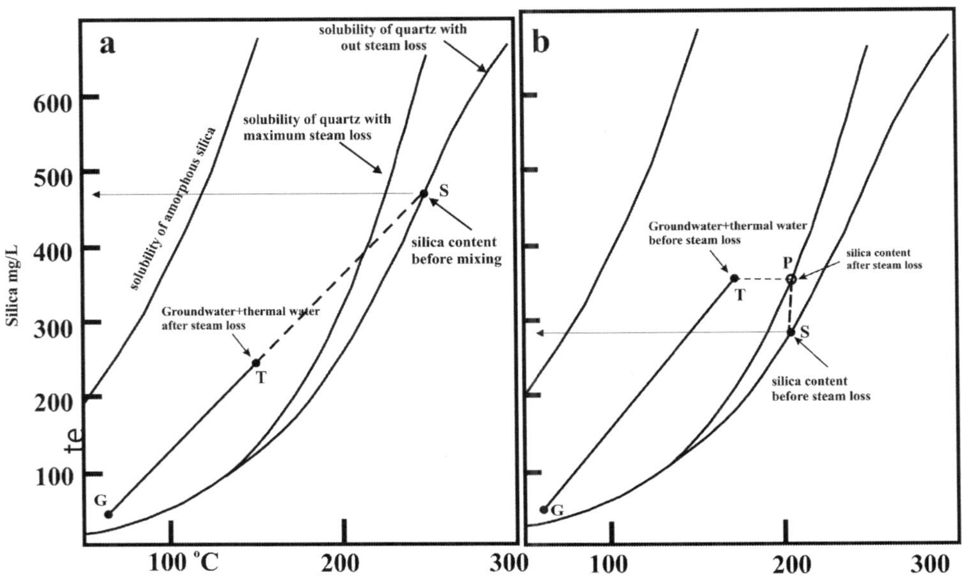

Figure 6.13 Estimation of silica in thermal waters mixing with groundwater. a) steam separation before mixing, b) steam separation after mixing (adapted from Chandrasekharam and Bundschuh, 2008).

prevailing at the near surface environment while reactions involving Na-K are not very sensitive and are not generally affected by processes that interact with the thermal waters during ascent (Giggenbach, 1988). Giggenbach tried to resolve this problem to a certain extent by combining both the reactions (temperatures) in a single diagram so that the deeper temperatures and near surface re-equilibrated temperatures could be visualised simultaneously. One such diagram proposed by Giggenbach (1988) is shown in Figure 6.14.

Figure 6.14 in combination with Figure 6.11 is generally used to make reasonable inference about the subsurface circulation patterns and reservoir temperatures of geothermal fields. The information obtained through Figure 6.14 and the temperatures obtained from various silica and cation geo-thermometers will provide a lot of information that is needed for planning the exploratory drill holes. Further, confidence in the subsurface rock water interaction processes and reservoir temperatures can be refined by using oxygen and hydrogen isotope signatures in the thermal fluids. The oxygen isotope signature in thermal waters is sensitive only at higher temperatures. Thus variations in oxygen isotope values provide significant information about geothermal systems.

6.2.1.4 *Oxygen and hydrogen isotopes*

Meteoric water makes up a major portion of geothermal systems. Volcanic terranes associated with subduction and non-subduction zones may contribute magmatic water to the circulating geothermal waters. The oxygen and hydrogen isotopes of geothermal

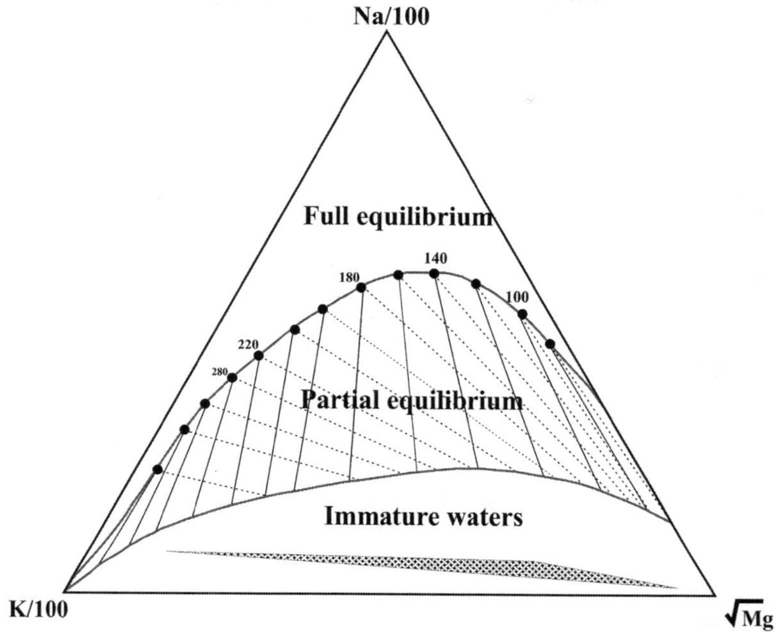

Figure 6.14 Na-K-Mg diagram showing the equilibrium fields and Na-K and K-Mg thermometers of Giggenbach (1988).

waters that have not undergone any isotope exchanges with surrounding rocks, have similarity with the meteoric waters and thus fall on the meteoric line in the $\delta^{18}O$ *vs* δD diagram of Craig (1961). Any deviation from this indicates environmental changes that these waters have undergone. Truesdell and Hulston (1980), Gat (1966), Giggenbach, (1992, 1998) and Nuti (1991) conducted extensive analyses of the behaviour of these two isotopes in thermal waters.

The general expression of oxygen and hydrogen isotopes is

$$\delta = 10^3[(R_s/R_{SMOW}) - 1]$$

$$10^3 + \delta = 10^3(R_s/R_{SMOW})$$

where s is the sample and SMOW is the standard, and R is either ($^{18}O/^{16}O$) or (D/H). Thus, when the water samples are enriched with ^{18}O and D with respect to SMOW, then the $\delta^{18}O$ and δD values will be positive and when the water vapour condenses and falls as rain, the rain will be enriched with ^{18}O and D with respect to the water remaining in the cloud. This makes the water vapour progressively more negative. Therefore, the $\delta^{18}O$ and δD values of the rain fall will also be negative since the original vapour is depleted in ^{18}O and D. The isotope data is generally represented by the $\delta^{18}O$ and δD diagram shown in Figure 6.15.

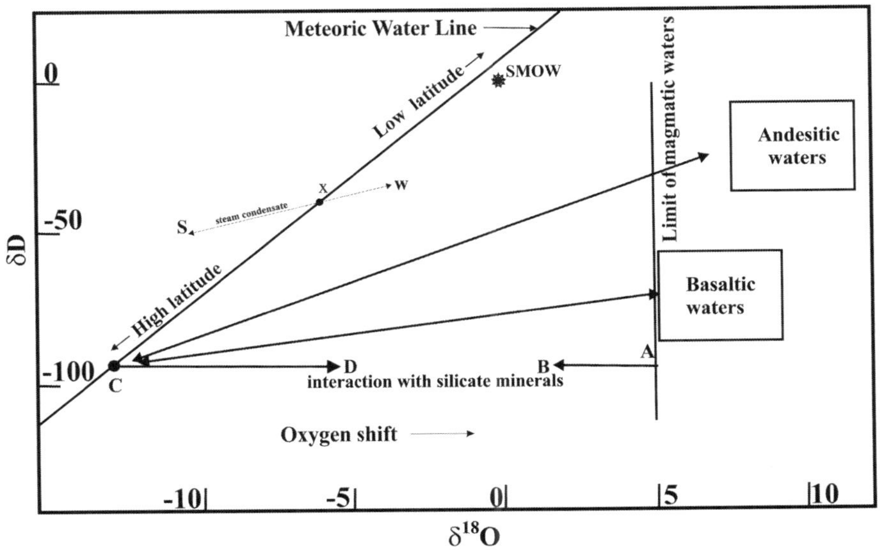

Figure 6.15 Oxygen and hydrogen isotopes behaviour in geothermal systems (adopted from Giggenbach, 1998).

The meteoric waters from higher latitudes and poles tend to have more negative isotope values while those from around equatorial regions have positive values. In general, precipitations around the globe define a line known as the meteoric water line or global meteoric line as defined by Craig (1961). This is the average value of isotopes in rain from all regions of the world. However, depending on local environmental variations, on a regional level, the isotope values show slight variation but the line defined by the isotope will be parallel to the global meteoric line. Geothermal waters that move away from this line indicate an exchange of these isotopes either from the rocks or from magmatic fluids. Since hydrogen is not a major component in silicate rocks, oxygen isotope exchanges between thermal waters and rocks only take place at higher temperatures (>220°C, Nuti, 1991). In this case, the thermal waters plot away from the meteoric line but parallel to the oxygen isotope axis in Figure 6.15. When magmatic water mixes with thermal waters, the resultant mixer falls on the tie line connecting the thermal water and magmatic waters; both oxygen and hydrogen isotopes show a shift depending on the composition of the magmatic water (Figure 6.15). The oxygen isotope exchange between water and rock takes place at high temperatures (220°C, Nuti, 1991). During the water and rock exchange reactions, while the thermal waters gain oxygen isotope and move towards a positive $\delta^{18}O$ value, in the donour mineral phase the isotope tends to move towards negative value. Thus in Figure 6.15 the plots of thermal waters and mineral phase move in the opposite directions. For example, when the thermal waters, due to the incorporation of an oxygen isotope, move from C to D in Figure 6.15, this isotope value in the mineral phase shifts from A to B. In the case of thermal systems associated with active volcanic terrain, steam condensate from the thermal waters, register more negative $\delta^{18}O$ and δD and move towards the left side of the meteoric line, for example X-S in Figure 6.15, while the thermal waters tend to move towards positive $\delta^{18}O$ and δD, for example X-W, in Figure 6.15. The higher the temperature of the isotope exchange, the larger is the oxygen isotope shift (Faure 1986, Nuti, 1991) in Figure 6.15.

6.3 GEOPHYSICAL EXPLORATION METHODS

Surface exploration methods (geological and geochemical) for geothermal resources have their own limitations since these methods cannot provide sufficient information about the subsurface factors controlling the geothermal systems. The main emphasis of exploration is to highlight factors that are sensitive to fluid temperature and fluid content and extract information about the subsurface structures that play a role in influencing the geothermal systems. Geophysical methods help in understanding the geothermal reservoir characteristics such as areal extent of the reservoir, depth, fractures controlling the fluid paths, geothermal gradients, and heat flow. Physical parameters like porosity of the reservoir rocks, permeability, temperature and pressure can be deciphered through a variety of geophysical methods.

The geophysical methods that are applied to geothermal exploration include seismic, electrical resistivity, potential (gravity and magnetic), temperature gradient and heat flow measurements. Electrical resistivity and electromagnetic methods (transient electromagnetic (TEM) and magneto telluric (MT)) are commonly applied in geothermal exploration. All geophysical methods follow a common procedure that includes 1) collection of data and 2) data processing and interpretation of the processed data. The processed data may sometimes enable an interpretation of the subsurface information and sometimes the processed data is utilised to evolve a computer model or a mathematical model to interpret the results.

6.3.1 Electrical resistivity method

In this method, electrical current is passed into the earth through current electrodes and the potential difference is measured at the surface through potential electrodes. This method is based on Ohm's law. According to Ohm's law $E = \rho j$ where E is the electrical field strength (V/m) and j is the current density (A/m^2) and ρ is the electrical resistivity (Ωm). Electrical resistivity can also be defined as the ratio of the potential difference ΔV (V) to the current I (A) across a material that is one metre long with a cross sectional area of $1\,m^2$ ($\rho = \Delta V/I$). The electrical properties of rocks provides important information like temperature, geothermal fluid characteristics (related to salinity), porosity and status of the rocks (weathered/un-weathered) which are directly related to the geothermal systems. Because the earth is not a homogenous body, resistivity surveys give apparent resistivity, ρ_a, which assumes as if the earth behaves like a homogenous body. ρ_a is converted to the true resistivity of the subsurface strata through modelling. There are three main types of resistivity surveys adopted in geothermal exploration: 1) direct current (DC) method, 2) transient electromagnetic method (TEM) and 3) Magneto telluric (MT) method. In the DC method, current is generated and passed into the earth through electrodes at the surface. The electrical field generated is measured at the surface. In the TEM method current is passed into the earth through a time

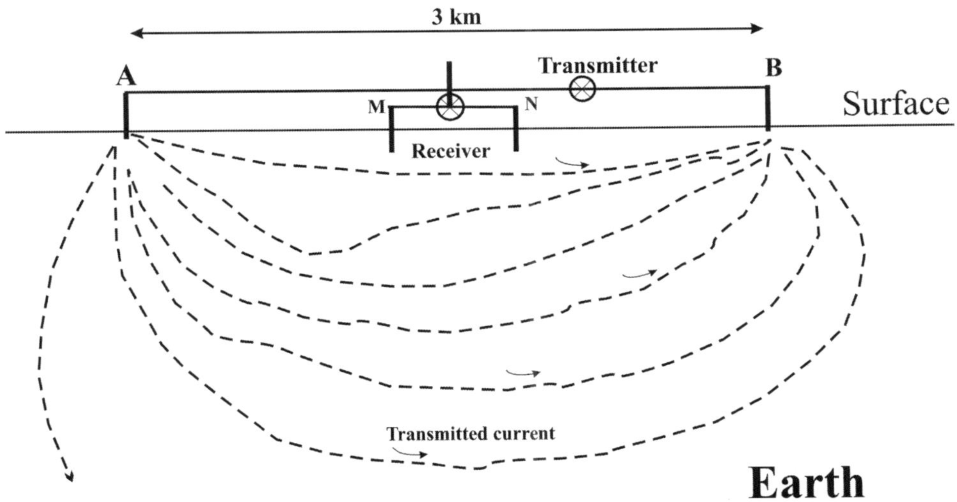

Figure 6.16 DC resistivity method with Schlumberger electrode configuration (adapted from Hersir, and Flovenz, 2013).

varying magnetic field from a controlled source. The decaying magnetic field is measured at the surface as a secondary magnetic field. In the MT method, current is passed into the earth by the time varying earth's magnetic field and the electromagnetic field is measured at the surface (Georgsson et al., 2005, Hersir and Flovenz, 2013, Kana et al., 2015).

6.3.1.1 DC method

In this method direct current (DC method) is passed into the ground using a pair of electrodes over the ground surface. This produces an electrical field which is measured as the potential difference between the electrodes; this is related to the resistivity of the underlying rock formation. Depending on the configuration of the electrodes, the DC sounding method is known as Schlumberger and Wenner method. The Schlumberger electrode configuration is commonly used in geothermal exploration. The Schlumberger electrode configuration is shown in Figure 6.16.

In Figure 6.16, the current (AB) and the potential electrodes (MN) are equidistant from the centre. The potential electrodes are kept close to the centre at a fixed distance while the current electrodes positioned away from the centre and moved away from the centre after each set of measurements. The distance between the current electrodes increases in a logarithmic scale. The distance between the current electrodes is typically 1 to 3 km in geothermal exploration surveys. The greater the distance between the current electrodes, the deeper is the current penetration (Kana et al., 2015)

$$\rho_a = \Delta V/I(S^2 - P^2)\Pi/2P = K(\Delta V/I)$$

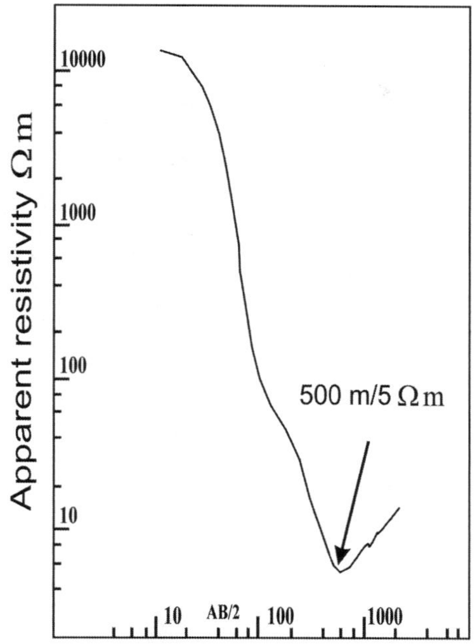

Figure 6.17 Graph showing the relationship between the apparent resistivity and half-electrode distance.

where S = AB/2, P = MN/2 and K is a geometrical factor, ΔV = potential difference, and I = current.

The apparent resistivity is plotted against AB/2 on a log-log scale as shown in Figure 6.17. The apparent resistivity curve can be inverted to obtain the information about the thickness of the subsurface strata and the resistivity of the rock formations. The high temperature/high conducting strata (geothermal reservoir?) here is located at a depth of about 500 m.

6.3.1.2 Magneto telluric method

The magneto telluric measurements permit the detection of resistivity anomalies that result due to the presence of faults, high temperature fluids, and reservoirs. These features are part of any geothermal system. A magneto telluric survey is a passive electromagnetic exploration technique very commonly adopted for geothermal exploration. This technique investigates the distribution of electrical conductivity in the earth's crust. The method works on the natural electromagnetic field of the earth. When the magnetic field reaches the surface, it induces electrical currents known as the telluric currents (eddy currents) in the conducting earth. These telluric currents induce a secondary magnetic field. The measured magnetic and electrical fields at the surface, give the electrical conductivity of the material below the surface. MT sounding survey can penetrate to about 10 km depth. A typical MT sounding survey set up is shown in Figure 6.18.

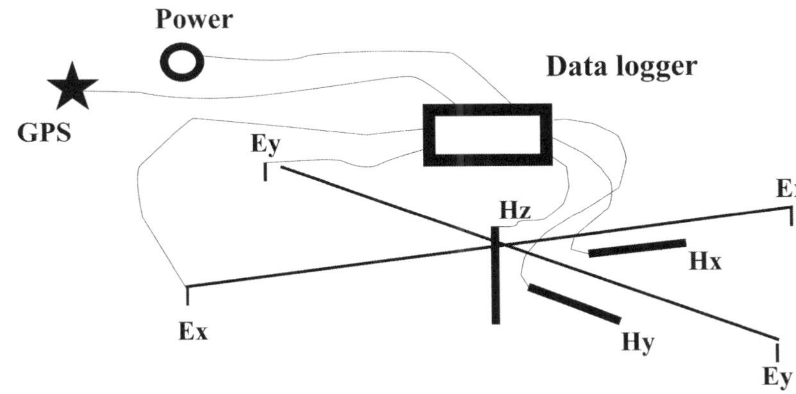

Figure 6.18 MT survey set up. Telluric currents: Ex-Ex Horizontal direction, Ey-Ey orthogonal direction. Magnetic field: Hx, Hy and Hz.

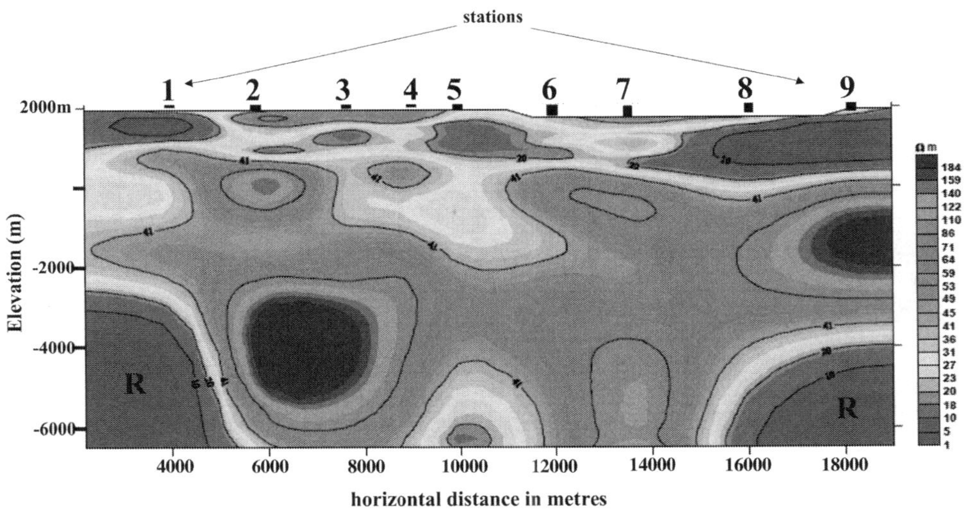

Figure 6.19 Schematic diagram showing typical 2D MT survey results of a high temperature geothermal province.

The telluric currents are measured on the surface in two horizontal and orthogonal directions (Figure 6.18) and the magnetic field is measured in three orthogonal directions.

The selection of a survey site over a potential geothermal area is very important to obtain reasonable data for interpretation. For example, if there are power lines crossing over the survey lines, they will interfere with the telluric currents and the data recorded will have considerable noise and this reduces the resolution of the data. The MT survey records time series of magnetic and electrical fields with wavelengths varying from 0.0025 s to 1,000 s or higher than 10,000 s. If seismic data for the area under MT survey is available then the interpretation will greatly improve. The earth's magnetic field is influenced by the ionospheric and magnetospheric currents that are caused by the solar flares or solar wind (plasma) or micropulsations. Lightning activity also causes sferics (high frequencies) near the equator that have frequencies >1 s. Generally, the MT measurements are made continuously for 20 hours with frequencies varying from 0.0025 to 1,000 s. The shallow structures are shown in higher frequencies while deeper structures are reflected in lower frequencies.

The recorded data is subjected to a Fourier transformation to transform the data from a time domain to a frequency domain. The apparent resistivity and phase, as a function of period, are calculated. The relationship between the electrical and magnetic fields can be established using the following equation:

$$\begin{bmatrix} E_x \\ E_y \end{bmatrix} = \begin{bmatrix} Z_{xx} & Z_{xy} \\ Z_{yx} & Z_{yy} \end{bmatrix} \begin{bmatrix} H_x \\ H_y \end{bmatrix}$$

E = electrical field vector, H = magnetic field vector, Z = impedance tensor that is related to the resistivity of the subsurface formation or structure. Using Z values, the apparent resistivity (ρ) and Phases (θ) for each period (T) can be calculated using the following equation:

$$\rho_{xy}(T) = 0.2T|Z_{xy}|^2; \ \theta_{xy} = \arg(Z_{xy})$$

$$\rho_{yx}(T) = 0.2T|Z_{yx}|^2; \ \theta_{yx} = \arg(Z_{yx})$$

For a homogeneous and 1D Earth, $Z_{xy} = -Z_{yx}$ and $Z_{xx} = Z_{yy} = 0$. Thus, parameters xy and yx, ρ_{xy} and ρ_{yx}, θ_{xy} and θ_{yx} are equal. The MT surveys can reach large depth penetration. However, the depth of penetration is controlled by the wavelength of the EM fields recorded and the resistivity of the different formations below the surface. The depth of penetration is the depth (δ) at which the EM field attenuates to e^{-1} of the surface amplitude.

$$\delta(T) \approx 500\sqrt{T}\rho(\mathrm{m})$$

where ρ is the average resistivity of the subsurface strata

A typical MT 2D profile is shown in Figure 6.19 and examples from the western Arabian shield are given in Chapter 4.

Chapter 7

Power generation systems

All developing countries have a stake in energy sustainability, but in the Asian region especially, large populations aspiring to greater prosperity will strongly test our ability to deliver energy sustainability. Collectively, we will need to distill all available wisdom on the policies, market structures, pricing arrangements and technologies that can lead us to our goals. These issues are also the ones, which pre-occupy industrialized countries.

Antonio Del Rosario, Welcome address at the World Energy Congress, 2004

7.1 TYPES OF GENERATION SYSTEMS

Once geothermal steam or thermal water is brought to the surface through a production well, a large part of geological and geophysical work are accomplished. Before the well is developed to generate power, the endurance of the reservoir or well is tested. The well, after development, is allowed to eject steam or water over a long period of time to note any changes that may occur in the flow rate, temperature or pressure (Figure 7.1). Once the well has passed this test, it goes to the power production stage.

Surface engineering technology plays a vital role in converting the steam or thermal water from the well to electrical power, until the spent water is pumped back into an injection well. Surface engineering technology depends on the product which comes out of the production well. If the geothermal system supplies only steam, then a single flash power plant design is adopted for running the turbine. If the well produces two phases (steam and water), both the phases are utilised for generating electricity. If the geothermal field produces only thermal water, then binary cycle technology is commonly adopted. A typical geothermal production well is shown in Figure 7.2.

Out of the three common technologies, any one of these technologies or a combination of them are commonly in use in all the geothermal fields of the world. If the geothermal field is producing only low-enthalpy liquid phase (<150°C, Chandrasekharam and Bundschuh, 2008), then binary cycle power generation technology is the only option to generate power and this, compared to the other two technologies, is the most widely used power generation technology in the world. These technologies are briefly discussed in the following section.

Figure 7.1 Geothermal well testing, Olkaria, Kenya (Photo by D. Chandrasekharam).

Figure 7.2 Geothermal production wells. (a) Miravallies, Costa Rica, (b) Wairakei, New Zealand, (c) details of a production well and (d) Olkaria, Kenya (Photo by D. Chandrasekharam).

Figure 7.3 Steam gathering pipes in a geothermal field (a) Olkaria, Kenya and (b) Wairakei, New Zealand (Photos by D. Chandrasekharam).

7.1.1 Single and double flash power generation systems

Several geothermal fields either produce only steam or water or they produce a mixture of the two. The steam producing wells are generally located over a wide area in a geothermal field. Steam from each well is gathered using a network of insulated pipes to a central receiving station and is directed to a turbine. An example of such a network of steam gathering pipes is shown in Figure 7.3. The steam pipes are often bent in a loop to allow for thermal expansion (Figure 7.3b). Some pipes are over 5 km long and expand to about 5 m.

In the case of a single flash system, the steam or steam and water mixture from the production well is directed to a cyclonic pressure vessel where the two phases with distinct density contrast get separated (Figures 7.4 and 7.5a). The separated steam is

Figure 7.4 Steam separator (a) Wairakei, New Zealand and (b) Olkaria, Kenya (Photo by D. Chandrasekharam).

then directed to run the turbine while the water is injected back into the aquifer through an injection well. If there are several production wells, then the cyclone separator can be placed at each well or near the power house, where all the steam and water are collected by a network of pipes (Figure 7.3). A typical cyclone steam separator is shown in Figure 7.4. The temperature-enthalpy state diagram of a single flash plant and a schematic diagram showing the design of a single flash power plant are shown in Figures 7.5a and b respectively.

As shown in Figure 7.2, a single well can generate about 4 to 6 MWe. After the steam separation in a steam separator, the separated liquid may still contain steam with sufficient enthalpy. In such cases, the hot water is flashed to extract additional steam in a flasher and the steam thus separated is directed to a low pressure turbine to generate electricity (Figure 7.6). The power plants of this type are called double flash power plants. The temperature-enthalpy state of a double flash power plant and a schematic assemblage of a double flash power plant are shown in Figure 7.6.

The steam generated from the flasher is of low pressure and is therefore used for generating power from a low pressure generator (Figure 7.6). Currently the turbines

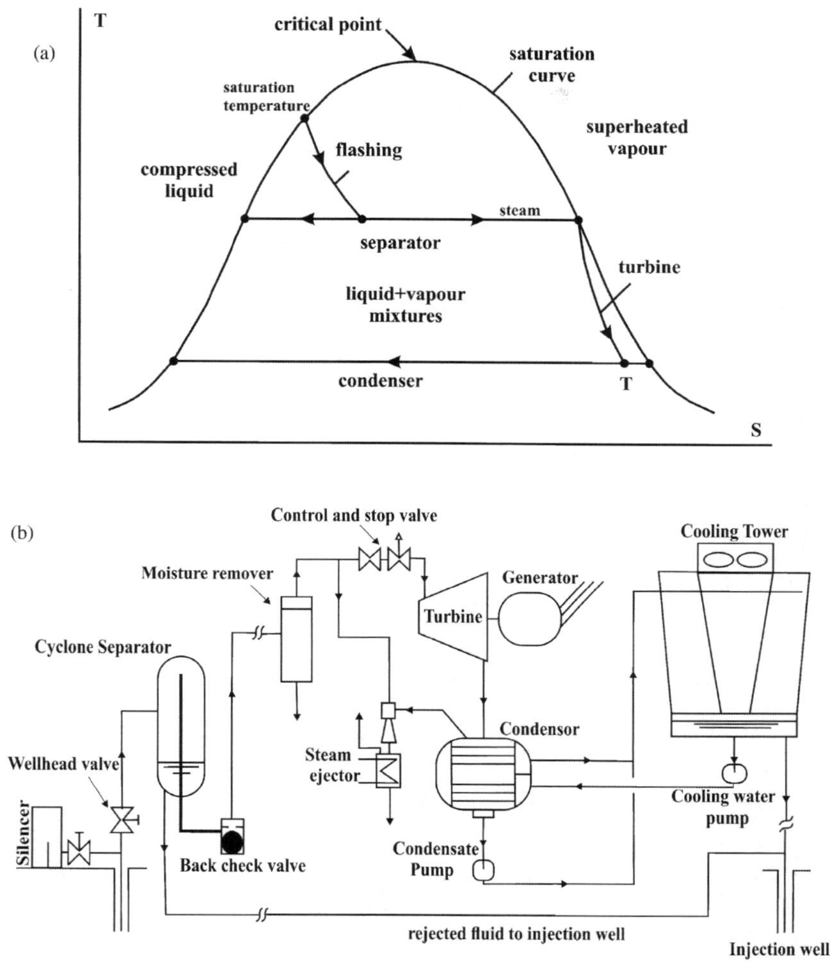

Figure 7.5 Temperature-enthalpy state diagram of a single flash plant (a) and a schematic diagram (b) showing the assemblage of a single flash geothermal power plant (adapted from Di Pippo, 2008).

are designed to receive both low and high pressure steam (dual admissible turbine) and thus reduce the capital cost of the power project. The advantage of a double flash steam plant is that it can extract more steam from the system and can generate additional power of the order of 15 to 25% (Di Pippo, 2008).

If the hot water is not fit for flashing, then the heat from the hot water is extracted through a heat exchanger to generate power using binary cycle technology (Figures 7.7 and 7.8). One such system is in operation in Wairakei, New Zealand. A twin binary cycle power plant receives rejected water (2800 tonnes/hr) with 130°C from 5 flash power plants (Figure 7.8). The rejected water pumped into reinjection wells is directed to the binary cycle power plants to generate 14.4 MWe and the water, after extracting

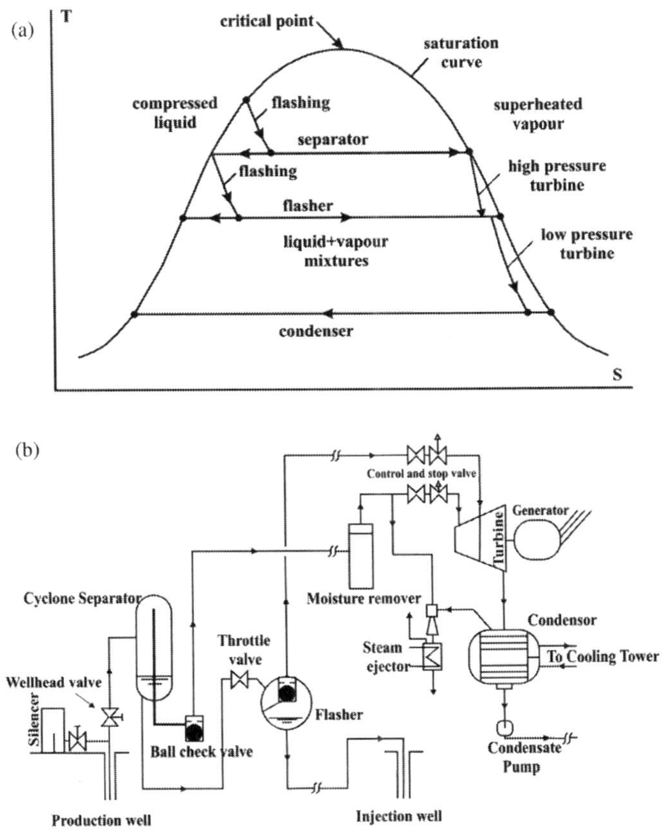

Figure 7.6 Temperature-enthalpy state diagram of a single flash power plant (a) and a schematic diagram (b) showing the assemblage of a double flash geothermal power plant (adapted from Di Pippo, 2008).

the heat (87°C), is directed to a prawn culture facility, before it is pumped into the injection wells.

7.1.2 Binary cycle power plant

The binary cycle power plant is the most common technology adopted because this system can receive geo-fluids with low temperatures, as low as about 76°C, and convert the heat to electricity (Chandrasekharam and Bundschuh, 2008). The binary cycle technology is also known as organic Rankine cycle (ORC) and is based on a principle similar to the power plants operated through fossil fuels or nuclear energy. The ORC system is very compact and can sometimes be easily shifted. The basic principle behind this technology is to extract the heat from the geothermal fluids through a heat exchanger and transfer the heat to a fluid with a low boiling point (known as the working fluid), generally an organic liquid. The general criteria for the selection of a working fluid are that the fluid should be non-corrosive, non-flammable and should

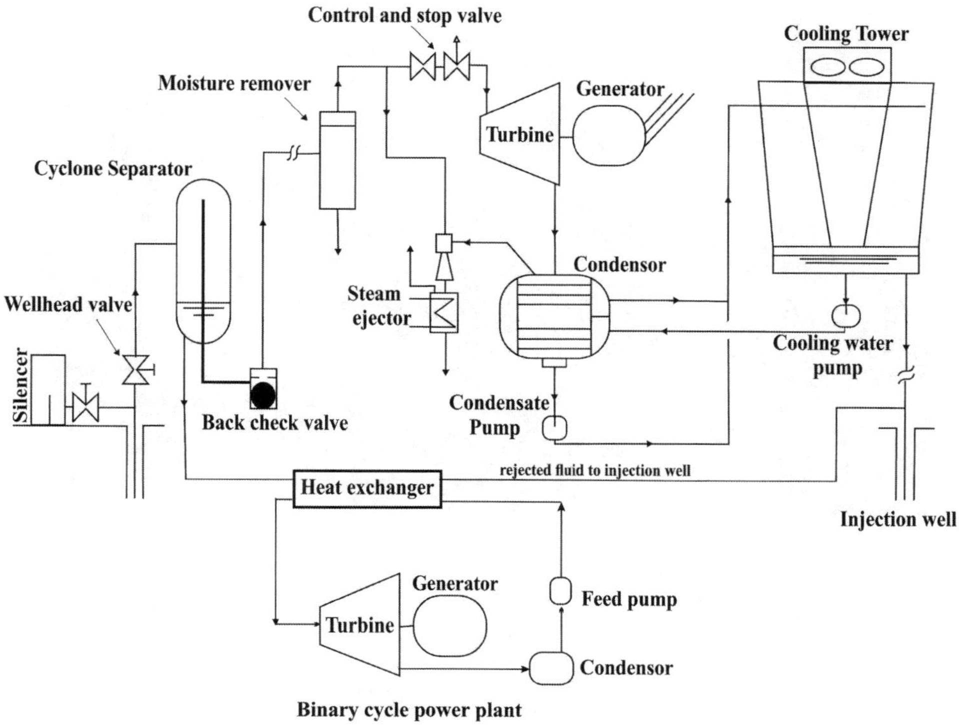

Figure 7.7 Schematic diagram of a combined single flash and binary cycle power plant.

not react or dissociate at the temperatures and pressures at which it is used. Many organic fluids do not meet these criteria and are flammable. Because of their low boiling point, these organic fluids are best suited to generate power from low-enthalpy geothermal fluids, with temperatures as low as 76°C. Further, as the organic fluids enter the turbines, low temperatures are obtained from the expansion of the vapours and thus the turbine chamber need not be under vacuum. The organic fluids that are most commonly used in the ORC technology are n-butane (BP: −0.5°C), i-butane (BP: −11°C) for thermal fluids with temperatures <200°C, and toluene (BP: 110°C) for thermal fluids >200°C). The organic fluid n-butane has higher efficiency relative to iso-butane (Vijayaraghavan and Goswamy, 2005). These organic fluids are, in fact, greenhouse gases (GHGs) and hence have the potential to cause environmental degradation. There are, therefore, also many constraints on the selection of these organic fluids. A list of organic fluids that are suitable for use as binary fluids in ORC power plants are listed in Table 7.1.

A typical schematic binary cycle geothermal power plant and the pressure-enthalpy diagram are shown in Figure 7.9.

All the listed binary fluids (Table 7.1) that are used in the ORC power plants are not environmentally-friendly and are highly toxic. Combining certain organic fluids has resulted in a new brand of binary fluids that are not toxic. These are R-11 and

Figure 7.8 A twin binary cycle power plant using 130°C reject water from flash power plants, Wairakei, New Zealand (Photo D. Chandrasekharam).

Table 7.1 Thermodynamic properties of certain organic fluids suitable for ORC cycle.

Fluid	T_c °C	P_c MPa	P_s @ 300 K MPa
Propane	96.9	4.23	0.99
i-butane	135.9	3.68	0.37
n-butane	150.8	3.72	0.26
i-pentane	187.8	3.41	0.097
n-pentane	193.9	3.24	0.07
Ammonia	133.6	11.67	1.06
Water	374.1	22.09	0.003

R-12 refrigerants that had been in use by the geothermal industry for a while until they were banned due their high ozone depletion potential (ODP) and global warming potential (GWP) by the IPCC (Intergovernmental panel on Climate Change).

7.1.2.1 Kalina cycle

The climate related issues associated with organic fluids in ORC were overcome in recent years by the Kalina cycle, which uses a mixture of ammonia and water that has a higher volatility, and hence a higher efficiency. compared to the conventional ORC plants (Vijayaraghavan and Goswamy, 2005). This technology was invented by a Russian scientist Dr Alexander Kalina in 1983 (Kalina, 1983). It is also known as the ammonia-water cycle and was used mainly to run engines. Subsequently this design was modified to suit several industries, especially to recover waste heat from, for

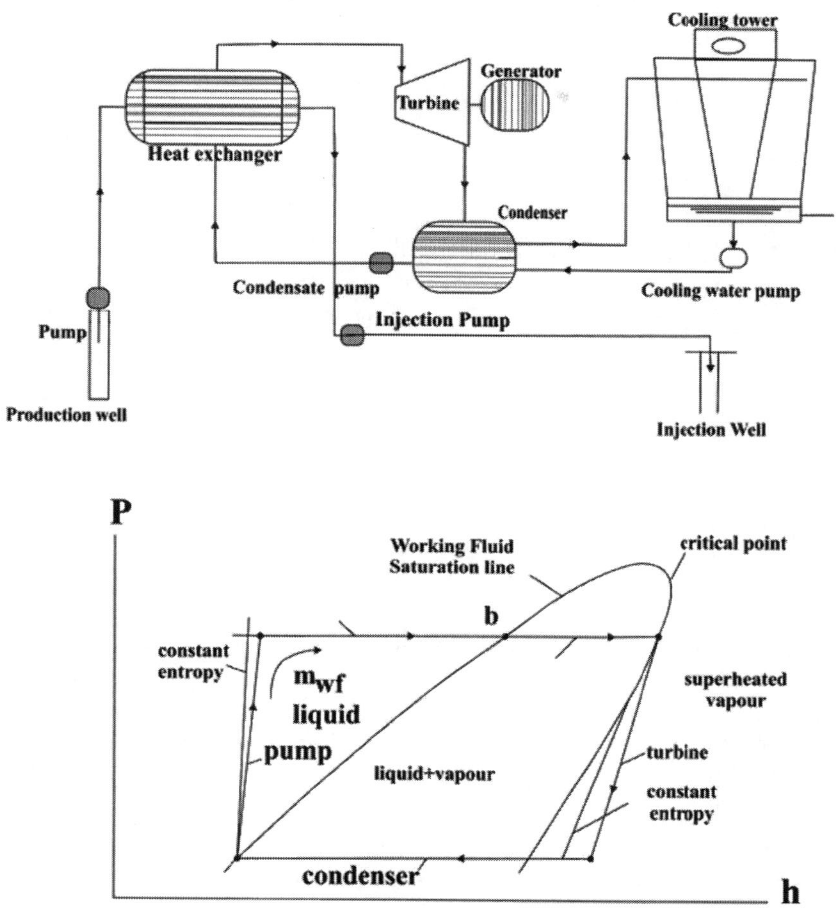

Figure 7.9 Schematic binary cycle geothermal power plant and pressure-enthalpy diagram of a binary cycle.

example, a cement factory, to generate electricity. This was further developed to suit the geothermal industry. Since ammonia and water have different boiling temperatures with near similar molecular weights (ammonia: 17 kg/kmol, water: 18 kg/kmol), both the liquids can be separated with ease. The ammonia-water mixture evaporates over a wide range of temperatures, depending on the ammonia to water ratio. Thus, this binary fluid is best suited for generating power from high and low-enthalpy geothermal sources (Kalina and Leibowitz, 1987). Ammonia is not a greenhouse gas and therefore its ODP and GWP are low. The economic advantage of the Kalina cycle over the ORC has been analysed by several researchers: Kalina et al., 1981, Kalina and Leibowitz, 1989, Lazzeri, 1997). This system is superior to the ORC in terms of power output, efficiency and heat transfer from geothermal fluids. For geothermal fluids above 110°C, the Kalina cycle's power output from geothermal fluids is higher, compared to the ORC (Figure 7.10).

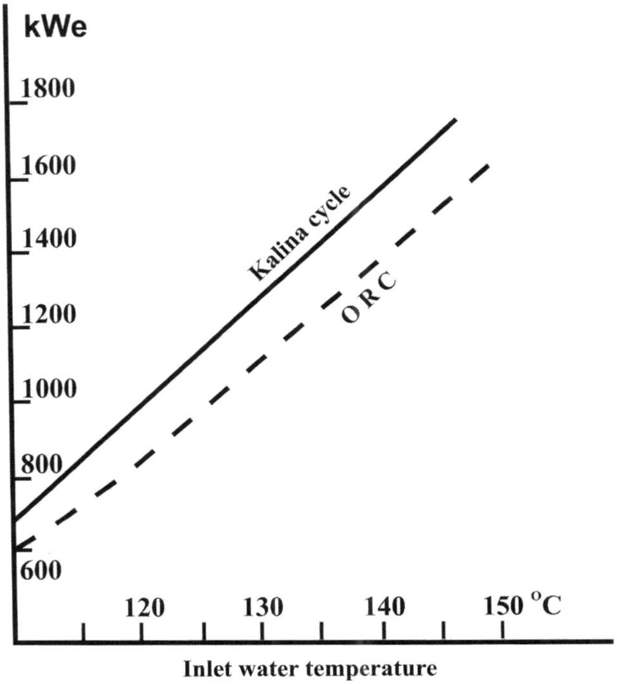

Figure 7.10 Electricity generation capacity of Kalina and ORC cycles (adapted from Chandrasekharam and Bundschuh, 2008).

Although the capital expenditure of the Kalina cycle technology is more than that of the ORC, this is compensated by the Kalina cycle's ability to generate more electricity compared to the ORC at a given inlet temperature (Figure 7.10) and its efficiency over the ORC (Figure 7.11). Although the heat exchanger in a Kalina cycle is bulky, this is compensated by its ability to transfer more heat compared to an ORC (Figure 7.12) and the fact that the turbine size is small in the Kalina cycle compared to the ORC and consequently the cost is less (Jonsson, 2003).

Because of this efficiency of the heat exchanger in transferring the heat from the thermal water to the secondary fluid (nearly 100% heat transfer), this system can generate 20 to 40% more power compared to the ORC at the same given inlet temperature of the thermal water (Figure 7.12). Overall, for the Kalina cycle, though little complex than the ORC, the investment cost is similar to the ORC based power plants. The plant efficiency, lower generation cost, reduced emissions risks (no GHGs are involved), and ability to transfer a high percentage of heat from the thermal fluid to the secondary fluid, make this technology more suitable for low as well as high-enthalpy geothermal sources relative to the ORC.

Kalina has developed two systems, one that is suitable for a high-enthalpy system (Kalex SG-4d) and a second one suitable for a low-enthalpy system (Kalex SG 2a). The performances of the Kalina and ORC binary systems, in terms of power production and efficiencies, utilising the same flow rate of thermal fluids from production well

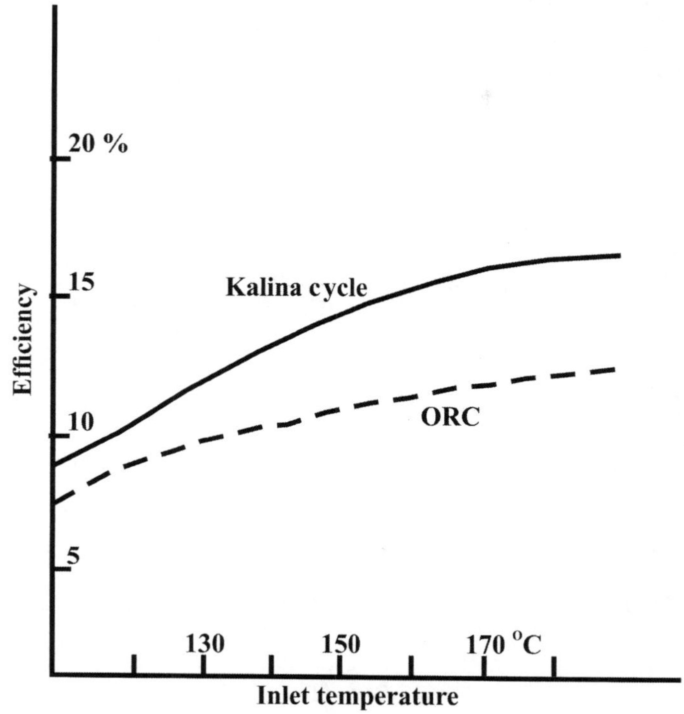

Figure 7.11 Efficiency of Kalina cycle and ORC (adapted from Valdimarsson and Eliasson, 2003, Chandrasekharam and Bundschuh, 2008).

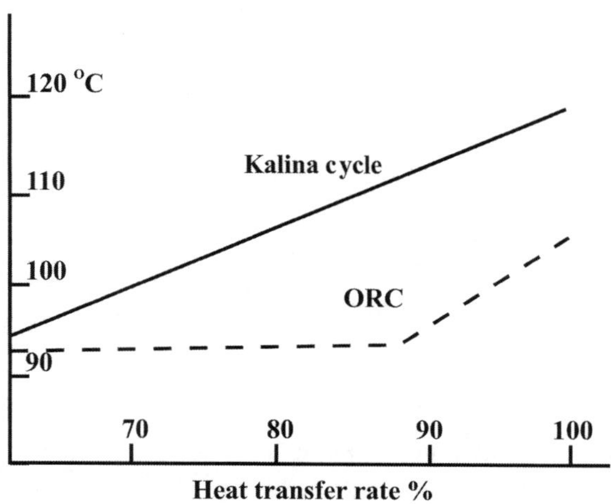

Figure 7.12 Heat transfer rate in Kalina and ORC systems (Kalina, 1984).

Table 7.2 Comparison of power output (MWe) for 1,000,000 lbs/hr flow rate between Kalina and ORC (DP: double pressure, SP: single pressure, adapted from Kalina, 2006).

In °C	*Out*	*Kalex SG-4d*	*Kalex SG 2a*	*ORC DP*	*ORC SP*
193	75	11.4		10.2	9.4
187	75	10.7		9.4	8.6
182	75	10		8.7	7.9
176	75	9.3		7.9	7.2
171	75	8.7		7.3	6.6
165	73	8		6.7	6
160	71	7.4		6.1	5.4
154	68		6.7	5.5	4.8
148	66		6.2	5	4.3
143	64		5.6	4.5	3.8
137	62		5	4	3.4
132	60		4.5		2.9
126	57		4		2.5
121	55		3.6		

Table 7.3 Efficiency (%) in power output between the Kalina cycle and the ORC under similar operating conditions (adapted from Kalina, 2006).

In °C	*Out*	*Kalex SG-4d*	*Kalex SG 2a*	*ORC DP*	*ORC SP*
193	75	18		16	14.8
187	75	17.8		15.6	14.3
182	75	17.5		15.1	13.8
176	75	17.1		14.6	13.3
171	75	16.9		14.2	12.8
165	73	16.2		13.5	12.1
160	71	15.2		12.8	11.4
154	68		14.8	12.1	10.9
148	66		14.1	11.4	10.4
143	64		13.4	11.5	9.9
137	62		12.6	10	9.5
132	60		11.9		9.1
126	57		11.1		8.7
121	55		10.4		

and ambient air temperature, have been calculated by Kalina (2006). The results are shown in Tables 7.1 and 7.2.

Under the current global warming scenario, it is critical when choosing a technology to evaluate its GHGs emission status and CO_2 emission reduction potential without sacrificing the socio-economic development of any country. In this respect, the Kalina cycle is superior to the ORC and will continue to be so as several countries around the world have large low-enthalpy resources that are lying untapped (Hettiarachchi et al., 2007, Chandrasekhar and Bundschuh, 2008).

Chapter 8

Direct applications

"The development of greenhouse agriculture and geothermal-based aquaculture in my country also demonstrates how sustainable energy can increase food production considerably, giving farmers and fishermen new ways to earn a living."

Ólafur Ragnar Grímsson, President of Iceland, 2011

There are several ways that geothermal energy can be utilised. In the previous chapters, power generation from geothermal heat was discussed in detail. Besides power generation, geothermal energy can be used directly for several applications. They include 1) space cooling and heating for individual homes and for district heating on commercial basis, and heat pumps, 2) dehydration and greenhouse cultivation, 3) aquaculture, 4) industrial processing, and 5) for balneology. Direct utilisation of geothermal energy is one of the oldest and most common forms of utilising geothermal energy (Lund and Boyd, 2015). The world installed capacity for direct utilisation of geothermal energy is 70329 MWt and the total annual energy use is 163287 GWh (Lund and Boyd, 2015). The direct utilisation of geothermal energy in the world over a period of 10 years, from 2005 to 2015 is shown in Table 8.1.

The world over, about 80% of the electricity is being utilised for space cooling, especially in tropical countries. This leaves a large volume of CO_2 in the atmosphere, which affects the global climate. CO_2 emissions can easily be mitigated if geothermal energy is used for air conditioning large commercial complexes, multi-storey apartment blocks and individual houses as well.

Table 8.1 Direct use of geothermal energy (MWt) during the past ten years (Source: Lund and Boyd, 2015).

MWt *Category*	*2015*	*2005*
Heat Pumps	49898	15384
Space heating + cooling	7916	4537
Greenhouse cultivation	1830	1404
Agriculture (drying)	161	157
Aquaculture	695	616
Industrial use	610	484
Batlung/Balneology	9140	5401

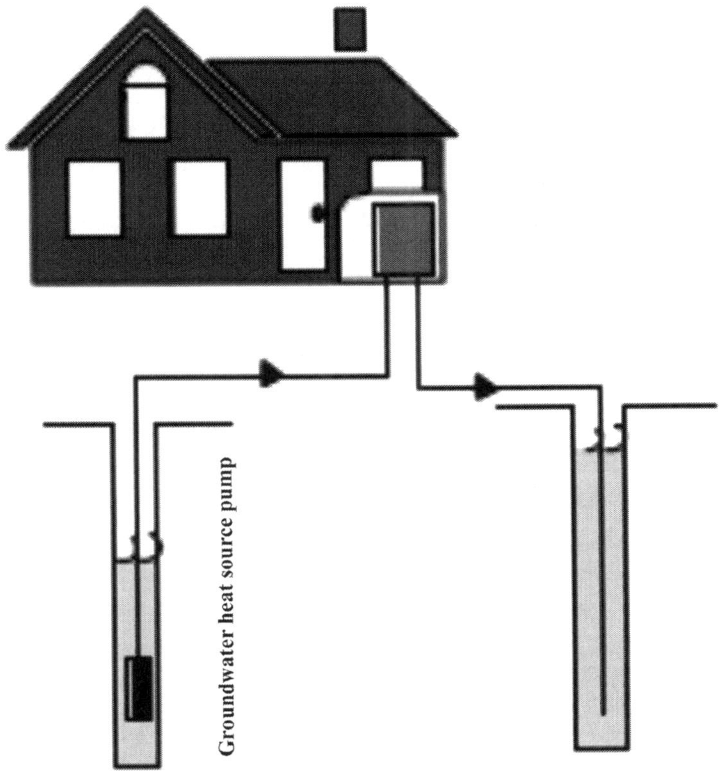

Figure 8.1 Open loop heat pump configuration.

8.1 THE GROUND SOURCE HEAT PUMP

Ground source heat pumps (GSHPs) are now very commonly used for heating and cooling buildings. The environmental impact of the heat pumps is minimal. The system needs a small input of electricity and, if this supply is from a renewable source, then the net output of CO_2 is negligible while if the supply is from conventional sources, then a small amount of CO_2 is emitted. GSHPs transfer heat to and from the earth to provide cooling and heating for homes and buildings. Thus the earth is a source and a sink of heat. The medium of the heat carrier or heat remover is a liquid, either groundwater or a refrigerant. The process is similar to a refrigerator where circulation is aided by a heat pump (compressor in the case of a refrigerator). During the summer months, heat is removed from the homes and transferred to the earth while in winter earth's heat is transferred to the homes. The system takes advantage of earth's inherent thermal behaviour- in summer the ground is cooler and in winter it is warmer.

The mechanism of transferring the heat from or to the earth is accomplished by long tubes, which can be placed horizontally or vertically. The system can be open or closed. In the case of an open system, groundwater is the main circulating fluid (Figure 8.1) while in a closed system, a refrigerant is the circulating fluid (Figure 8.2).

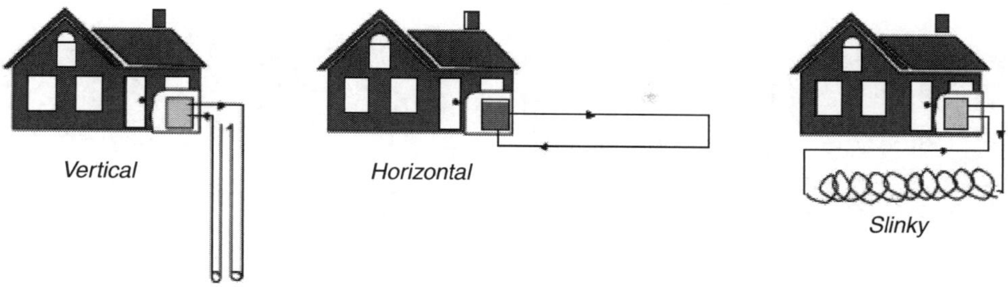

Figure 8.2 Closed loop heat pump configuration.

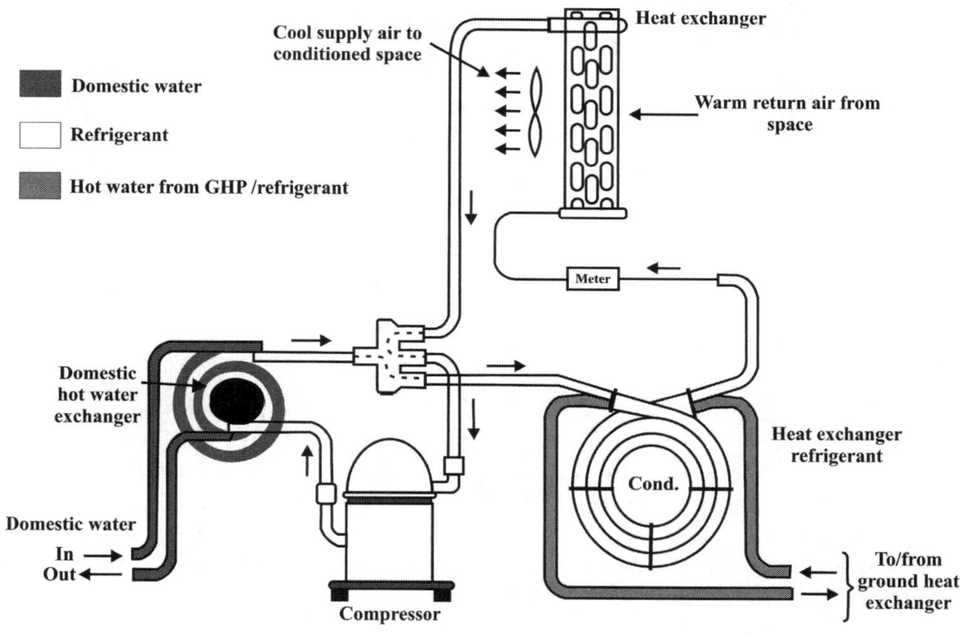

Figure 8.3 Cooling cycle of a GHP.

The closed system can be installed at any place and is easier to install. The heat exchanger, a loop of coil, can be placed horizontally or vertically below the ground. If there is a constraint in space to lay the loop horizontally, a vertical design is an option (Figure 8.2).

Compared to the horizontal loop configuration, the vertical loop has the advantage of providing a constant mean earth's temperature controlled by a geothermal gradient. The horizontal loop, as it is placed at shallow levels, is affected by the surface temperature of the earth which is affected by rain, solar radiation and wind. Schematic diagrams of the GSHPs are shown in Figures 8.3 (cooling cycle) and 8.4 (heating cycle).

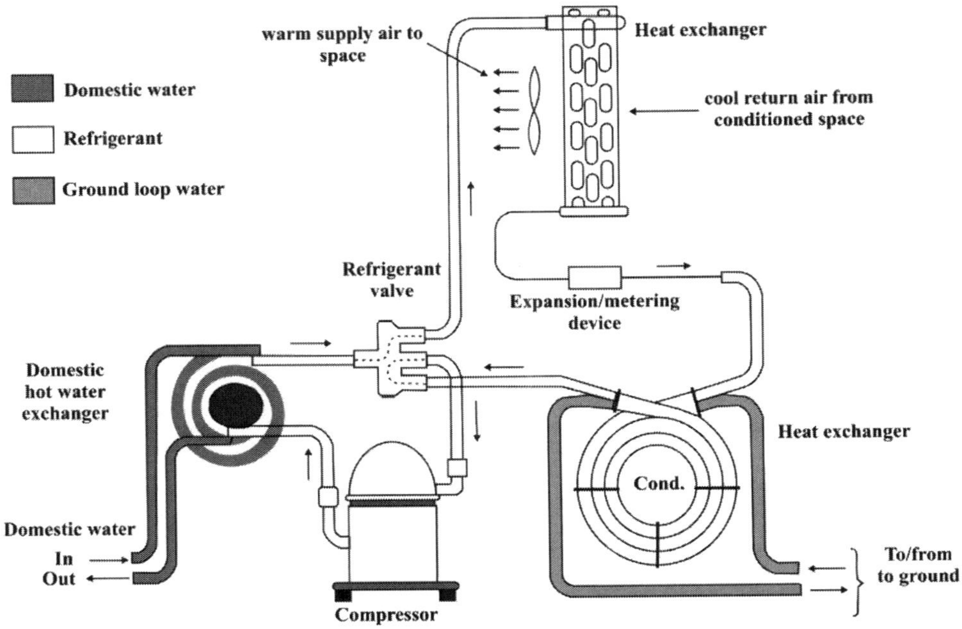

Figure 8.4 Heating cycle.

8.2 GREENHOUSE CULTIVATION

Plants need a specific temperature for growth. For example, plants that grow in tropical climates may not be able to sustain growth in cold climate regions. Greenhouse cultivation has become a global industry for growing vegetables, fruits and flowers. The advantage of a greenhouse is that the plant growth is not restricted to any season, hence the required food can be grown at any time of the year for domestic consumption as well as for commercial export. A typical greenhouse, cultivating bananas, flowers and vegetables and roses is shown in Figure 8.5.

Figure 8.5 demonstrates how effectively geothermal water can be utilised for greenhouse cultivation irrespective of the air temperatures. Several countries have taken this as an industry and several entrepreneurs have started agricultural businesses using geothermal based greenhouse technology. Greenhouse cultivation in cold climate regions is very attractive and is cost effective and the products can be sold to the local communities at low cost. Countries like Kenya have large geothermal based rose cultivation for commercial purposes in Olkaria. One bore well drilled to a depth of 1.6 km producing 51 T/hr of hot water and steam is completely dedicated for rose cultivation and the flowers are exported worldwide. The growth of the plants, development of fruits and vegetables depends on the temperature maintained in the greenhouse. For example, as shown in Figure 8.6, different vegetables have different maximum growth temperatures.

Geothermal greenhouses can reduce the operational cost and the products can be grown on a commercial scale. Since the humidity is controlled, a plant's growth will be healthy and free from diseases (Schmitt, 1981).

Figure 8.5 Greenhouse cultivation using geothermal water. a) banana, b) gourd, c) leafy vegetables, d) sweet lime and e) roses (photos by D. Chandrasekharam).

Greenhouse heating can be carried out either by 1) circulating air over finned coil heat exchangers carrying hot water in plastic tubes running along the length of the greenhouse. This maintains uniform heat throughout the length of the greenhouse, 2) pipes carrying hot water over the floor of the greenhouse, 3) finned units located along the walls or below the benches or 4) a combination of the above three. The most economical and efficient greenhouses are a large structure covering a large area, say about 36 × 110 m constructed with fibreglass with furrow connected gables (Lund, 1996). Heating would be through a combination of fan coils connected in series with a network of horizontal pipes installed outside walls and under the benches (Lund, 1996). A storage tank to store geothermal waters is necessary to meet any peak demand. About 6.3 L of 60–80°C water will be necessary for peak heating. The cost of contraction is about US$ 54–108 (1996 rate, Lund, 1996). A typical geothermal supported greenhouse is shown in Figure 8.7.

The greenhouse heating system is crop specific. The crop grown, the common diseases that attack the crops, humidity requirement of the crop and circulation air to control leaf mildew are some of the factors that guide the construction of greenhouses. Crops and plants that grow in the tropics and subtropics may require high humidity and

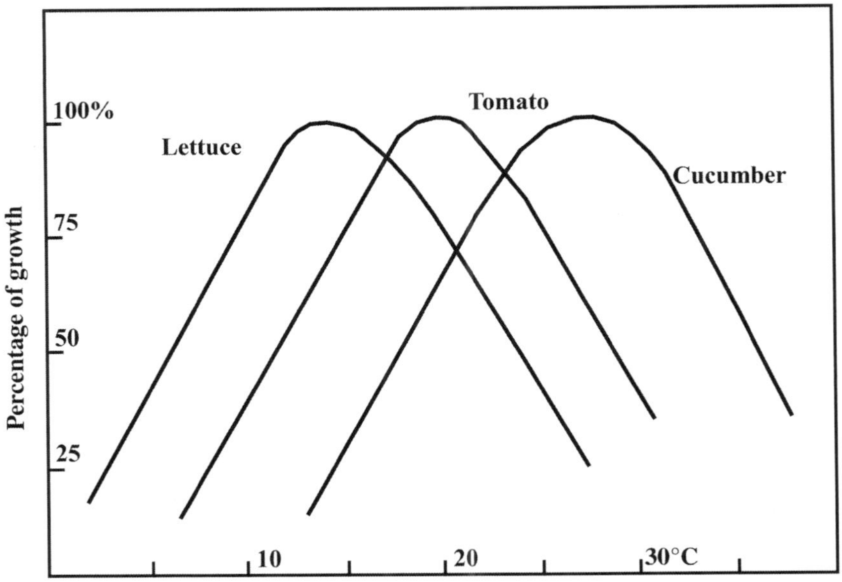

Figure 8.6 Growth response to greenhouse temperatures by different vegetables (Source: Lund, 1996, 2002).

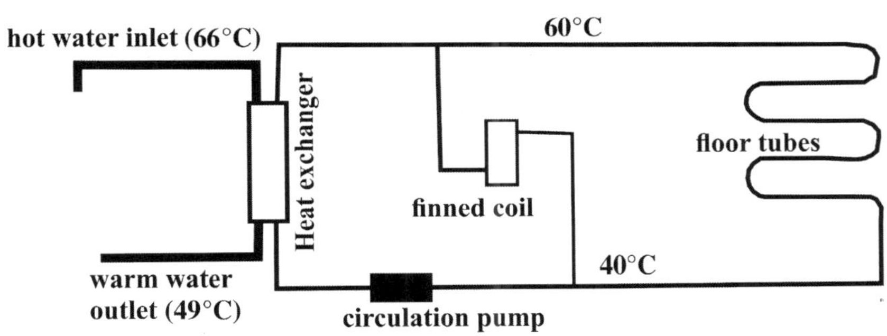

Figure 8.7 Schematic design of geothermal water circulation system in a typical greenhouse (Lund, 1996).

high soil temperatures. Certain flowering plants may need shading to control blooming. Detailed specifications for growing crops and flowering plants in a greenhouse can be found in Lund (1996, 2002).

8.3 DEHYDRATION

The application of dehydration technology in agricultural products using geothermal energy has shown tremendous growth in the world. Between 2005 and 2015 there has only been a slight increase in the energy utilised for this purpose in the world

Table 8.2 Cost comparison between binary power plant and dehydration plant (Source: Lund and Lienau, 2002).

	Binary power plant	*Onion dehydration plant*
Capital cost US$	50 million	15 million
Gross revenue US$	11 million	30 million
Resource rquirements	760 L/s	76 L/s
Number of employees	15	75

(about 4 MWt, see Table 8.1) compared to the period between 2000 and 2005. In 2000 the energy used for dehydration in the world was only 67 MWt while in 2015 it was 167 MWt. This huge jump is apparently due to the food industries becoming aware of this process. More than fifteen countries use geothermal energy for drying various grains, vegetables and fruits. Countries like Iceland (seaweed), USA (onions), Serbia (wheat and cereals), El Salvadar (banana), Guatemala and Mexico (fruits), New Zealand (alfalfa), Philippines (coconut pulp), Indonesia (timber drying) have established geothermal based agro-industries. China is emerging as the top user of geothermal energy for the dehydration of agricultural products.

8.3.1 Onion dehydration

Many vegetables contain nearly more than 50% water and hence cannot be stored for a long time after harvesting. One such vegetable is onions which contains nearly 83% water. Onion dehydration requires only low temperature air; thermal water with temperatures between 38 and 100°C is very suitable for this industry. In fact, several geothermal industries use a cascading technique where the thermal water after generating power in a binary cycle plant is directed for such use. This gives additional revenue to the industry. In fact, countries like Guatemala are keen to utilise geothermal waters for dehydration rather than for generating power using binary cycle plants (Chandrasekharam and Bundschuh, 2002).

About a decade ago, the Geothermal Development Association in Reno, Nevada calculated the costs of operating a binary cycle power plant and an onion dehydration unit using a geothermal source with the following specifications: 1) Temperature of the thermal water 150°C, 2) 20 MWe binary power plant (90%) availability, 3) unit cost of power (kWh) generated from the power plant is 7 US cents, *versus* onion dehydration plant using the same resources operating for 10 months, with two dryers in operation, 13.6 million kg of onions are processed (dried) and sold at US$ 2.20 per kg (whole sale price). The cost comparison is given as in Table 8.2.

It is apparent that the investment in the dehydration plant is quite attractive relative to the power plant. The land requirement and logistics support is less cumbersome in the case of the dehydration plant. It can also provide opportunities for entrepreneurs and for the creation of more employment. There are several products that can be dehydrated and a few are listed in Table 8.3 together with their production costs.

In addition to vegetables, the moisture from food grains can also be controlled by using geothermal heat. Removing moisture from food grains like wheat, barley,

Table 8.3 Cost comparison of dehydration of fruit using conventional (coal) and geothermal energy sources (adapted from Chandrasekharam, 2001).

Product	*Capacity (kg)*	*Time (h)*	*Geothermal US$*	*Conventional US$*
Pineapple	817	18	1.3	76
Apple (slice)	771	16	13	69
Apple (cubes)	907	16	13	69
Banana	817	24	21	86

oats, corn (Lund and Lienau, 2002) and fruits like dates (Kulkarni et al., 2008) will increase the food storage time, food availability during off season cultivation and contain the cost of agricultural products during inflation (Lund and Lienau, 2002). In the case of dates, dehydrated dates with desirable sensory qualities can be developed by processing immature dates. Blanching the dates with hot water (~96°C) for 15 min and then drying them by blowing hot air (60°C) will make the dates appear glossy and soften their texture which is in great demand (Kulkarni et al., 2008). A typical grain dryer is shown in Figure 8.8.

It is estimated that to dehydrate about 150 million tonnes/year of vegetables, 200 kWe or 12,26,400 kWh is required. Since coal based thermal power plants are used in many countries for dehydration, using this amount of electricity will emit 1.3×10^6 kg of CO_2. This amount can be mitigated by using geothermal energy sources (Chandrasekharam, 2001).

A typical dehydration unit using geothermal heat is shown in Figure 8.9. These units are portable and can be located near the farms. The products can be packed and exported directly to their destination after dehydration. The advantage here is that the plants will be located in rural areas, thus decongesting the urban areas and saving diesel. The cost is lower and most importantly the CO_2 emissions are reduced.

8.3.2 Milk pasteurisation

Milk pasteurisation using geothermal sources is in use, for example in Klamath Falls, Oregon USA and in Iceland.

Geothermal water with a temperature of 87°C and flow rate of 119 L/s has been utilised for the pasteurisation of milk in Klamath Falls. The geothermal water at 87°C was passed through a plate exchanger where the milk was heated from 3 to 71°C. The milk is then passed through the homogeniser and then passed through a second section of the heat exchanger where the milk is heated for 15 seconds at 77°C. Then the milk is culled to 12°C and finally chilled to 3°C by cold water. The advantage of this process is that the milk retains its flavour and its shelf life is enhanced (Lund, 1996). The milk is processed at the rate of 0.8 L/s (Lund, 1996).

8.3.3 Spas and balneology

A large number of thermal springs are used at tourist resorts for recreation and at health resorts for curing skin and bone ailments. Low temperature waters are either

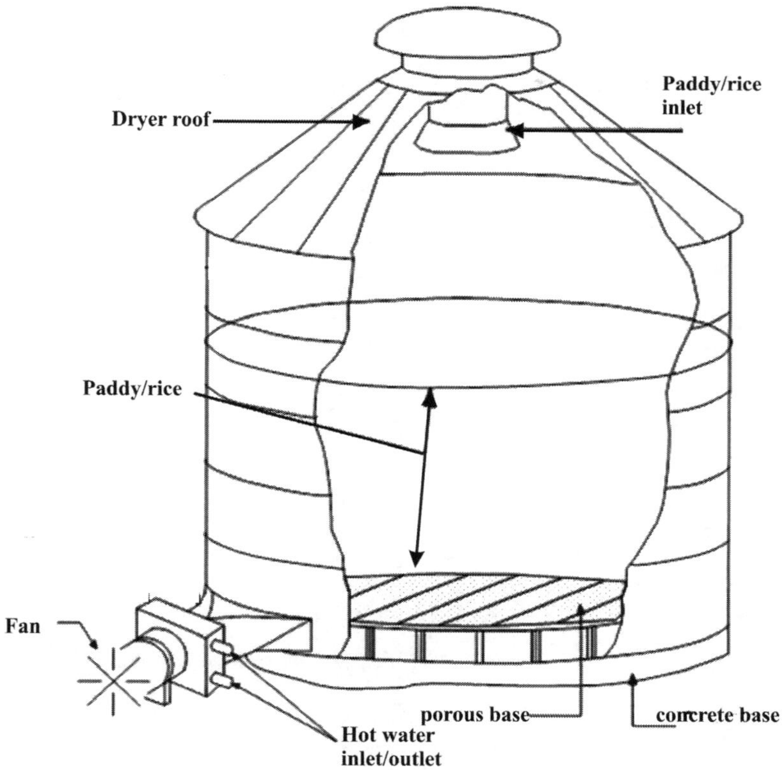

Figure 8.8 Grain dryer for removing the moisture from the food grains using geothermal heat (Source: Lund, 1996).

directly used or mixed with cold water to suit body temperatures. The tail water after generating power in a binary cycle power plant, before injecting into the injection well can also be utilised for bathing and swimming, as is being practised in Iceland. There are several examples and sites across the world where such facilities exist and can be found in the literature. The best compiled stories related to this aspect of direct use can be found in Cataldi et al. (1999).

8.4 DIRECT UTILISATION OF GEOTHERMAL SOURCES BY COUNTRIES AROUND THE RED SEA

8.4.1 Egypt

Direct utilisation of geothermal waters is not new to Egypt. Ancient Egyptians used warm water for domestic and medical purposes. Currently the country is utilising low temperature thermal waters for district heating and fish farming as well as greenhouse cultivation. Swimming pools using geothermal waters are constructed along the Suez Gulf coast for the tourists. The oases of Bahariya and Dhakla (Figure 4.2) are

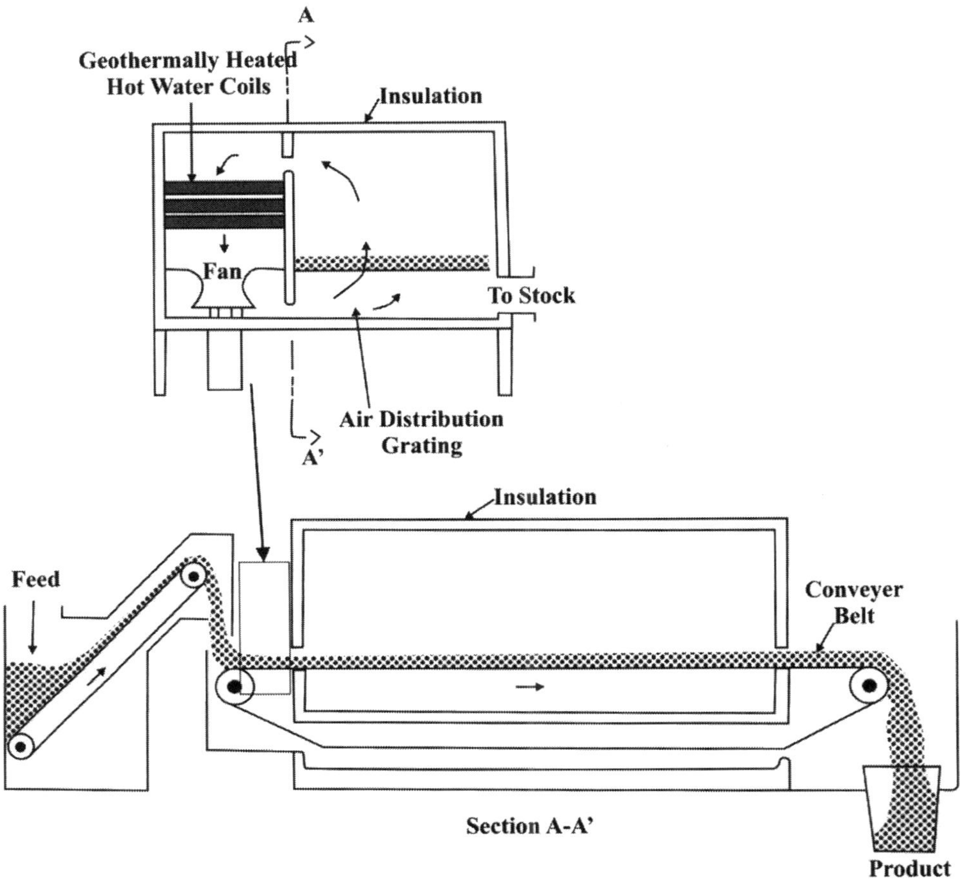

Figure 8.9 Schematic diagram showing a typical dehydration plant (adapted from Gundmundsson and Lund, 1987).

using thermal water for greenhouse cultivation. As per the published report (Lashin, 2015), swimming and bathing facility (4 MWt), a greenhouse (1 MWt), one space heating facility (0.3 MWt) one district heating facility (1.5 MWt) are in operation using geothermal waters as on date (Lashin, 2015). Most of Egypt's agricultural activity is centred around the Nile and a large part of the delta is being utilised for food production. The nearest geothermal source to the agricultural hub of Egypt is Ain Sukhna, which can be used for direct applications.

8.4.2 Eritrea

To date there are no direct application activities in this country. Eritrea is one of the least developed countries in the world with a low per capita income. Two thirds of the population depend on rain-fed agricultural activity and hence the country's food security remains a grave concern to the government. Although the country has sizable geothermal resources, these sources have yet to be developed. Once developed the country can come out of its socio-economic problems and develop into a developed country with proper energy and food security (section 2.2.2). Because of its dependency on imported fuel, any cost in fuel fluctuation affects the country's growth. Geothermal development will stabilise the country's economy and lift the country above the poverty line. Geothermal sources can reduce fuel imports, reduce foreign debts, and provide sustainable food and water security to the country. Eritrea is investing in three priority areas 1) food security, 2) infrastructure, and 3) human resources development (WB, 2014). As long as the country depends on imported fuel, it is difficult for it to achieve these three important goals.

8.4.3 Ethiopia

Compared to the above two countries, Ethiopia's economy is growing and the per capita growth is expected to grow to $698 from $452 (WB, 2012). Ethiopia, though currently not involved in direct geothermal use, considering the climate and geology, it can develop geothermal based greenhouse cultivation of roses, like Kenya, in and around Tendaho. Once developed the country can come out of poverty and secure its food and energy sources. These two vital sources are required for the country at present to rise above the poverty line. The rural population, in particular, can benefit from the geothermal resources around Tendaho and Dubti being developed completely. Cereals are the main crop cultivated. Ethiopia can utilise its geothermal energy for direct application like dehydration and greenhouse cultivation. These activities can be factored into the power generation projects where the tail water after electricity generation can be utilised for activities such as those mentioned above, before the waters are pumped back into the aquifer.

The geothermal waters from the sites located in the rift valley (southern part of Tendaho, Figure 4.10) are being utilised for swimming and bathing (2.2 MWt) in the hotels in Addis Ababa (Lund and Boyd, 2015). Since Ethiopia has considerable water sources, its agricultural activities are centred along the Awash River. The country's primary focus is to reduce oil imports and reduce CO_2 emissions by replacing biomass fuel in the rural areas with geothermal energy thereby mitigating health related issues (caused in the rural areas where biomass is used for cooking that pollute the kitchens) and enhancing the socio-economic status of the rural communities.

8.4.4 Djibouti

Djibouti has large geothermal potential that is underdeveloped. This country depends 100% on imported fuel even though it has very high geothermal energy potential. The geothermal development around Lake Asal and Lake Abhe can provide the necessary infrastructure, food, and water security.

8.4.5 Republic of Yemen

The thermal springs located in Damt, Dhamar (Figure 4.23) and those occurring along the escarpment near the Tihama plain are traditionally used for bathing and swimming. The reported thermal use is about 1 MWt (Lund, 2015). The country's

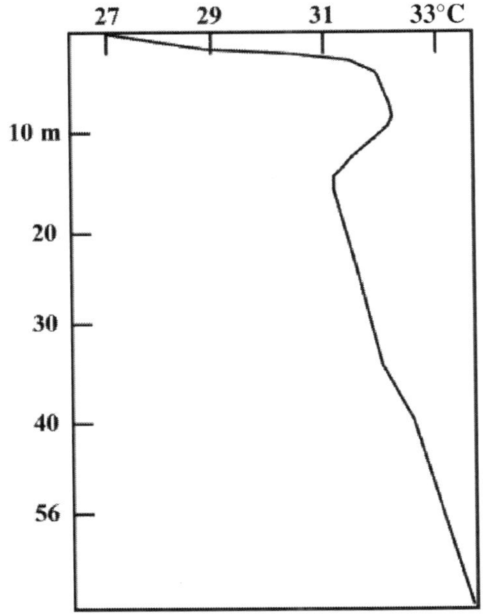

Figure 8.10 Shallow temperature gradient measurement in a bore hole in Riyadh (adapted from Sharqawy et al., 2009).

prime priority is to be energy independent. Once the geothermal power sector develops, the country's economy will grow opening opportunities for investment in industry and agriculture. The country has sufficiently large geothermal reserves (section 4) that needs to be utilised to boost the economy.

8.4.6 Saudi Arabia

The thermal springs around Jizan and Al Lith and Mecca are being utilised as spa and bathing centres. About 44 MWt of thermal water heat is being utilised for bathing and recreation purposes at Al Khouba hot springs in Jizan (Lund and Boyd, 2015). The Ministry of Tourism and the local municipality recently drilled a shallow well to tap into hot water. This water is used for medical therapy, spas and recreation. Thermal water is also used in animal farming (Lashin et al., 2015). Saudi Arabia has enormous hydrothermal and EGS potential (section 4) that can be used for a variety of direct applications discussed above. Among all the countries around the Red Sea, Saudi Arabia experiences extreme weather conditions with temperatures varying from 3 to 45°C. Over the past decade Saudi Arabia experienced an increase in air temperature by about 0.7°C as a result of high CO_2 emissions from its energy sector (Almazroui et al., 2012). Over 80% of the electricity generated from fossil fuel based energy sources is consumed for space cooling (air conditioning of commercial establishments, homes and offices (Chapter 3). At present 190 TWh of electricity is being consumed for space cooling of commercial and residential buildings to tackle harsh summers when the temperature soars beyond 57°C (Chandrasekharam et al., 2014). The world over 50,583 MWt (121,696 GWh/yr) of ground heat is being utilised with an annual increase of 12.3% (Lund et al., 2010). Germany and Japan are the leaders in utilising geothermal energy for space cooling and heating and other direct applications, followed by China. Usage of heat pumps showed a significant jump from 5,275 in 2000 to 35,236 in 2006 (Lund et al., 2010). Saudi Arabia too has an excellent opportunity to implement GHP technology to save 357,000 Gg of CO_2 emission. In a recent investigation, ground thermal conductivity has been investigated (Sharqawy et al., 2009) in the eastern part of Saudi Arabia (Figure 8.10).

Saudi Arabia is the world second largest date producer and its annual production is nearly 1.1 million tonnes of dates and the area covered by date palms is around 151,000 hectares, about 5% percent of this is exported. Due to stiff competition from other date producing countries like Tunisia, Saudi Arabia has to produce best dates and preserve them for export (El Juhany, 2010). Preservation of dates is an important business in the agricultural industry and in the competitive market, using conventional fuels to dehydrate the dates will yield less profit. If geothermal energy is utilised for dehydration, the product can be sold at low cost and the industry in Saudi Arabia can capture the world market (Chandrasekharam, 2001). The industry can earn additional revenue by saving CO_2 emissions and through carbon trading.

Chapter 9

Enhanced geothermal systems

We dedicate this book to those who labored over many years to take the hot dry rock concept from simply a novel idea to a proven reality. Their imagination, creativity, long-term commitment, and hard work led to the outstanding technical achievements that are described in detail herein. Those achievements have laid a solid foundation for the development of HDR geothermal energy as a major energy resource for the 21st century and beyond.

D.W. Brown, D.V. Duchane, G. Heiken, V.T. Thomas
Mining the earth's heat: Hot dry rock geothermal energy-2012

9.1 THE CONCEPT AND EARLY STAGES OF DEVELOPMENT

Hydrothermal energy sources are site specific and their occurrences are controlled by geological and tectonic configurations as discussed in Chapter 4. But unlike the hydrothermal systems, engineered geothermal systems (EGSs) are unlimited and present anywhere on earth. An engineered geothermal systems (EGS) has enormous potential to supply primary energy using heat mining technology. Heat mining technology is developed to extract and utilise the earth's thermal energy. Like the Sun's energy, earth's heat is unlimited. If the earth's heat supply stops, then there will not be any soul on this earth asking for energy! This is true for solar energy as well. EGS technologies have been tested at a number of sites around the world and the technology has improved since the 70s when the first hot dry rock (HDR) project was launched. The vision of the then Director of Los Alamos National Laboratory, Norris Bradbury (1945–70) has lead mankind to source unlimited energy from any part of the earth. His open invitation to scientists to come up with new ideas is now bearing fruit (Brown et al., 2012). HDR have different names such as man-made geothermal systems, enhanced geothermal systems and EGS. The principle behind them is what has been designed and developed in the 70s: to create a network of fractures at some depth where the rock is hot, and circulate water and extract the heat to generate electricity. Present EGS work is based on the concept patented by Potter et al. (1974) which describes the formation of a fully engineered geothermal reservoir in a hot crystalline rock such as granite, by creating a network of fractures through hydro-fracturing and extracting the heat by circulating water. The research was sponsored by the US Department of Energy until the technology to extract the heat from the rock was practical and economical. EGS technology is now mature and recently a road map was drawn

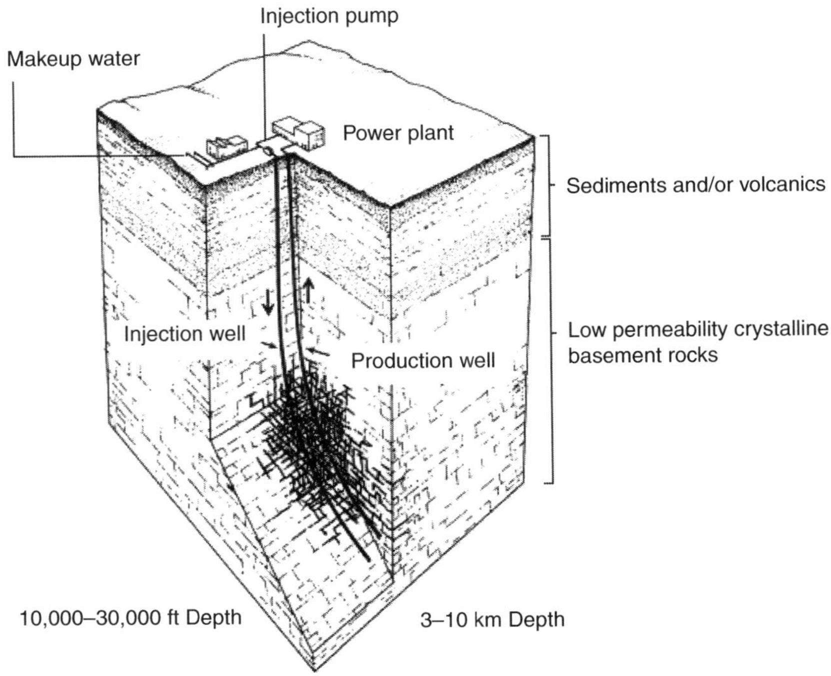

Figure 9.1 Schematic diagram of a HDR system.

to empower all countries to get an unlimited supply of clean and carbon free electricity (MIT, 2006). A schematic diagram showing an HDR system is shown in Figure 9.1.

EGS, like hydrothermal energy sources, can provide continuous baseload power with minimum impact on the environment. The land requirement for hydrothermal as well as EGS, unlike other renewables, is very small (Table 9.1). The CO_2 emissions are negligible and, in fact, in future EGS may use CO_2 as the working fluid to extract heat from the granite reservoir (Pruess, 2006). Once this technology matures, both coal based and EGSs can co-generate surplus electricity to the world. For example, it is estimated that, for the United States of America, the total resources base is 3×10^{15} MWh and extractable power is about 55×10^{12} MWh (MIT, 2006). The current consumption of electricity (from all sources) annually (until the end of December 2014, EIA, 2015) was 31×10^{12} MWh.

9.1.1 Fenton Hill experiment

With the full support given by Norris Bradbury, the Director of LASL, a group of scientists in the 70s, concluded that it should be possible to extract earth's heat by drilling two holes in the deeper crystalline rocks and connecting them by a network of fractures and circulating pressurised water through the drill holes. The initial task was to find an area where the rocks, at a reasonable depth, are sufficiently hot, with low permeability and should be able to support the construction of a loop to circulate water

Table 9.1 Land area required for EGS with different production capacities (adapted from MIT, 2006).

Plant size in MWe	*Surface area required for the power plaut (km^2)*	*Rock volume below the surface (km^2)*
25	1.0	1.5
50	1.4	2.7
75	1.8	3.9
100	2.1	5.0

through the induced fractures. Granites qualify as candidates satisfying the conditions. Moreover, due to the presence of radioactive elements like uranium, thorium and potassium, these rocks (natural nuclear reactors) have the ability to generate heat due to the decay of these elements. This heat, in addition to the heat conducted by the earth's interior, provides an excellent energy source. Due to their location (deeper depth), covered by soil and other non-conducting material above them, the heat generated by these rocks is stored in them. This heat is the largest supply of usable energy, which is available at no cost. In the case of other energy sources like gas, coal, oil and biofuel, the energy sources have to be bought at a cost. This energy is clean and no waste is generated, unlike other fuels. It also does not pollute the air. The land requirement of HDR is minimum compared to any other energy source. With the help of directional drilling, a large area of the granite below the surface can be reached. The Valles Caldera, a ring fault structure in New Mexico, was the site for this experiment. Surface manifestations of the presence of a thermal anomaly below the surface are indicated by several thermal springs. The basement rocks, below the 60,000 years old volcanic cover, lie at a depth of about 2 to 4 km.

The basement rocks are a highly jointed Precambrian crystalline complex containing granites and gneisses. The geothermal gradient of this site is about 64°C/km (Brown et al., 2012), greater than the average global value of 30°C/km. In the first phase, a 2,900 m deep bore well was drilled in granite where the temperature recorded was 197°C in the year 1974, four years after the project was conceived. After creating a network of fractures through hydro-fracturing, a second well was drilled establishing good connectivity between the first bore well. Between 1978 and 1980 several flow circulation tests were conducted. The area of the heat exchanger was found to be insufficient to extract the heat needed to generate power. Further, hydro- fracturing was carried out to enlarge the fracture network and subsequent flow tests indicate good heat extraction from the granite. Thus Phase I of the project established the credentials of the idea and increased confidence amongst scientists in the credibility of HDR projects in the future. A detailed Phase I Fenton Hill project was documented by Smith (1995). During the second phase of the project, a directional bore well to a depth of 4,390 m was drilled with the bottom hole temperature reaching 327°C. Subsequent flow tests established the credibility of the HDR project at Fenton Hill. Following this phase, several flow tests were conducted to assess the ability of the HDR system to increase the power output within a short period of time, in case a need arose in the real situation. The load-following experiment (LFE) proved that the HDR reservoir is capable of producing more power on demand. This is a significant achievement. Whenever there is a necessity, the HDR reservoir is capable of supplying more power (for example during peak operation of industries) within a short period of time. Thus, geothermal sources are the only energy sources that can supply baseload power, but can also supply more power on demand within a short period of time (Brown et al., 2012).

The Fenton Hill experimental project was a technological breakthrough which triggered interest in EGS by many countries. The major findings of the Fenton Hill project are 1) drilling to depth >5 km in hard crystalline rocks that are hot can be achieved, 2) hydro-fracturing the low permeability rocks to create a fracture network is possible, 3) by mapping the micro seismic activity, it is possible to connect two drill holes by the fracture network, 4) water can be circulated for prolonged period of time and the heat can be extracted, 5) directional drilling is possible to minimise the surface land footprint of the EGSs establishment, 6) well bore impedance can be managed, 7) data collected were used to develop models to understand the heat transfer and circulation behaviour of the EGS reservoir, 8) created a platform to develop high temperature tools and equipment that can be used in the bore wells to map the pre-existing fractures, monitor the pressure, temperatures and flow rates at greater depths.

After the Fenton Hill experimental HDR project, several field studies were conducted worldwide. EGS technology has advanced over the last thirty years and now the technology is very mature and generating power using this technology is possible by creating a fractures network in granites at depths of about 5 km. More than 2 km^3 rock volume can be stimulated. Issues like flow short circuit, need for high injection pressures, water losses, rock water interaction processes, and most importantly, induced seismicity, which are major concerns associated with EGS technology, are resolvable and manageable (MIT, 2006). In the case of high grade regions, i.e. regions where granites have extremely high heat generating capacity (thereby generating high heat flow and geothermal gradients) and occur at shallower levels, insulated by non-conducting thermal layers, the drilling depths can be brought down and consequently the cost of a project (Chandrasekharam and Chandrasekhar, 2008, 2010b et al., 2015a,b,d, Chandrasekhar and Chandrasekharam, 2008, Omenda et al., 2012, Lashin et al., 2014a). In fact, some of the hydrothermal systems in the world are driven by high heat generating granites (Lashin et al., 2014a, Chandrasekharam et al., 2015, Chandrasekhar and Chandrasekharam, 2009).

9.2 COMMERCIAL AND R&D EGS PROJECTS

After the Fenton Hill HDR experimental project, several countries started EGS R&D projects and some became commercial. The commercial and R&D EGS projects undertaken by different countries are shown in Table 9.2. Some of the project details are described in the following sections.

9.2.1 Rosemanowes HDR project

Located towards SW of London, this project started more or less during the second phase of the Fenton Hill project in 1977 (Table 9.2). The chosen site was Rosemanowes Quarry in Cornwall, with a target to drill in the Carnmenellis granite, with the Camborne School of Mines.

The Carnmenellis granite is part of a granite batholith which is porphyrytic at the surface and tends to be equigranular at depth. This batholith is exposed to the surface

Table 9.2 Commercial and R&D EGS projects by different countries (adapted from MIT, 2006, Chandrasekharam et al., 2015).

Start date	*Project site*	*Country*	*Well depth m*	*Rock type*	*BHT °C*	*Flow rate L/s*	*Reservoir Volume km³*	*Remarks*
1977	Rosemanowes	UK	2600	granite	100	15	1	R&D
1982	Ogachi	Japan	1027	granodiorite	230	not known		R&D
1987	Soultz	France	5093	granite	200	3.5	2.5	Commercial
1989	Hijori	Japan	2151	volcanic	225	17		R&D
1999	Cooper Basin	Australia	4325	granite	250	13	2.5	Commercial
2005	Paralan	Australia	4003	granite	171	6		R&D
2009	Geneys, Hannover	Germany	3900	Sandstone	160	7		R&D
2009	St Gallen	Switzerland	4450	Shell Lst	150	not known		R&D
2010	Newberry	USA	3066	volcanic	315	not known		Commercial

without any sedimentary cover. The heat flow measured at the surface of this batholith is about 120 mW/m^2 with a geothermal gradient of 35°C/km. The joints exposed at the surface extend to 2.5 km depth (Parker, 1999). The main aim of this project was not to produce power, but to conduct a large scale mechanical experiment to address some of the problems related to the hydraulic stimulation of natural fractures networks (Batchelor, 1982), reservoir development for commercial system, and application of new circulation fluids and sealing of fractures etc. The granite is characterised by vertical joints developed due to regional strike-slip faults. Thus, the mine works characterised the rock to a depth of 1 km. This is the reason why this site was chosen for this experiment. Further, this region has a normal geothermal gradient of 30°C/km. The targets for reservoir development (with a temperature of 210°C) are 1) production rate of the fluids to be 75 kg/s, 2) reservoir impedance would be 0.1 MPa, 3) water loss to be less than 10% and 4) cooling of the granite mass would be at the rate of 1°C/year over 25 years. These specifications amount to creating a reservoir volume in the granite of 300 million m^3. To create this volume the fracture area should be 5 million m^2 to produce 10 million m^2 of heat transfer area (Parker, 1999). In 1980, in phase II of the development programme, two bore wells (RH 11 and RH 12) were drilled to 2 km depth and the bottom hole temperature registered was 79°C. During the reservoir development programme 30,000 m^3 of water were pumped for fracture stimulation at a flow rate of 100 L/s maintaining a well head pressure of 140 bars. Water losses were registered at 70% due to a downward propagation of the fractures. The downward propagation of the fractures was tracked below RH 12 by another bore well (RH 15). A fracture stimulation using medium viscosity gel successfully connected RH 12 and RH 15 (Figure 9.2).

9.2.2 Ogachi HDR project

The Ogachi HDR site is associated with hydrothermal systems represented by thermal springs. Shallow initial test wells indicated a 300 m thick volcanic flow over hot granite (200°C) with low permeability. This project was the first of its kind where the

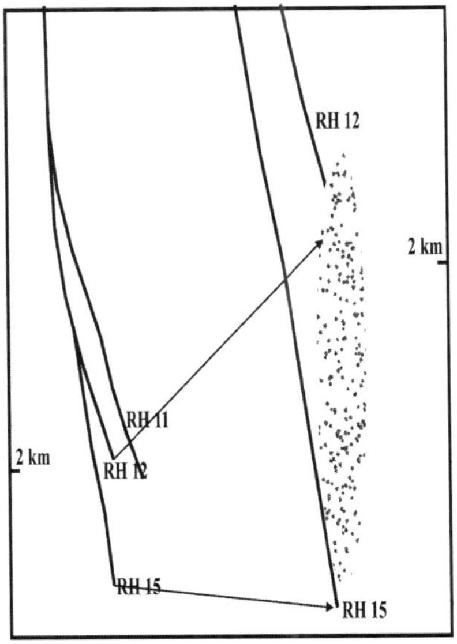

Figure 9.2 Drill hole configuration in Rosemanowes' HDR project, Cornwall, UK (adapted from Parker, 1999).

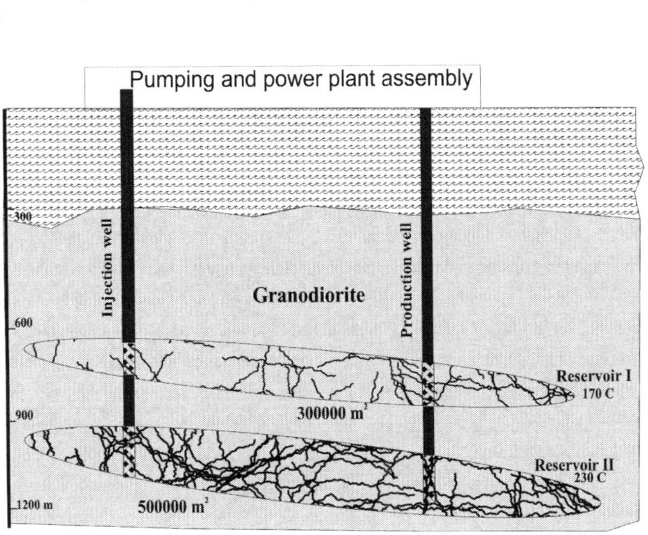

Figure 9.3 Schematic diagram showing the Ogachi twin HDR reservoir (adapted from Hori et al., 1999).

world's first two-layer reservoir in granite was proposed and drilled (Hori et al., 1999). The area of the upper reservoir was 300,000 m^2 and the lower one was 500,000 m^2. A schematic diagram of the reservoir is shown in Figure 9.3.

The casing reamer and sand plug (CRSP), a new technology, was tested and implemented in the Ogachi project to create a two-layer reservoir in granite. The Ogachi

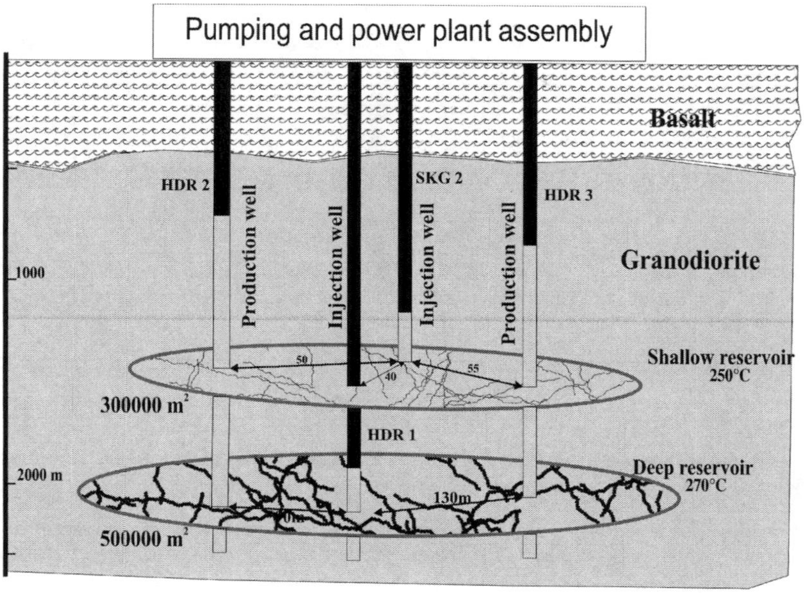

Figure 9.4 Conceptual diagram showing the Hijiori HDR wells (adapted from Tenma et al., 2001).

HDR project demonstrated new technology to create two-tier reservoir in granite. The fracture propagation and reservoir volume estimation was carried out by audio-frequency Magneto Telluric and Electromagnetic method (AE). A high temperature spinner and thermometer (HTST) was developed to measure flow rates under high temperature conditions. As per the cost estimate of power, the cost of a 75 MWe project is 18 yen per kWh, while a 240 MWe power plant would be 12.7 yen for a kWh. This higher production capacity rate is similar to the cost of a hydro-electric power plant in Japan (Hori et al., 1999). Thus, the Ogachi HDR project developed new technology to create a twin reservoir in granite. This project contributed to the advancement of technologies to the world's HDR community. These technologies are also cost effective.

9.2.3 Hijiori HDR project

The project site is located on the southern edge of the Hijiori caldera of the Gassan volcano which erupted 10,000 years ago. These volcanic flows lie over fractured granodiorite basement rock. The granodiorite was targeted to create an HDR reservoir. Thus are has a very high geothermal gradient due to the recent volcanic activity. Although the regional stress regime is compressive, the edge of the caldera exhibits a complex stress pattern. The experience gained from the Fenton Hill HDR project, specifically the new energy development organisation (NEDO) who participated in the Fenton Hill HDR project, planned to initiate the HDR project at Hijiori. The Hijiori HDR project also aimed at creating a twin reservoir (Figure 9.4). One injection well (SKG 2) and two production wells (HDR2, HDR3) were drilled to a depth of 1,800 m;

one well HDR1 was drilled to 2,151 m. The well HDR1 was later used for injecting the deeper reservoir. Natural fractures were encountered by all the drill holes. The distance between the wells was not large and varied between 40 and 55 m. The temperature recorded was 225°C at 1,500 m and 250°C at 1,800 m (Tenma et al., 2001).

Hydraulic fracturing and stimulation experiments were carried out for 30 days to test the circulation and connectivity of the wells. Steam and hot water were produced from HDR2 and HDR3. A total of 44,500 m^3 of water was injected and only 13,000 m^3 water were produced with 70% loss of water during the circulation test. In the case of the test with HDR1, out of the 51,500 m^3 of injected water only 26,000 m^3 could be produced as steam and hot water and the recovery was 50%, better than earlier test. After long circulation tests, the wells were able to produce 5 kg/s of hot water and steam at 163°C (HDR2a) and 4 kg/s of steam and water at 172°C. The total thermal power production achieved in this project was 8 MWt and a pilot binary cycle power plant was able to generate 130 kWe (Tenma et al., 1998, 2008, MIT, 2006).

The most important outcomes of the project are: 1) change of direction of the stress field away from the well. This is a setback in predicting the propagation of the fractures. This could be controlled by modifying the pressure in the wells, 2) it is easier to connect the fractures by wells rather than connecting wells through hydraulic fracturing, 3) stimulation at low pressures and for longer periods gives good results, and 4) it is possible to control and optimise the net thermal output of a complex HDR reservoir by manipulating the injection depths and injection flow rates.

9.2.4 Soultz HDR project

The Soultz HDR site is located on the western edge of the Rhine Graben. The project aimed to create a reservoir in granite at 5 km depth by hydro-fracturing and circulating water to produce hot water and steam which can be used for generating power. The granite was covered by a 1,400 m thick sediments. GPK1 was the first well drilled in 1987 to a depth of about 2,000 m. The temperature recorded at that depth was 140°C.

GPK1 was stimulated in 1991 creating a fracture volume of 10,000 m^3. Subsequently, this well was deepened to 3,590 m and the temperature recorded at this depth was 168°C. Since this GPK1 was under a tensional regime, it was assumed that large scale injection of fluids was not necessary to create fracture network connectivity. However, in practice fluid injection was necessary to accomplish fracture connectivity (Baria et al., 1998). Three wells GPK2, GPK3 and GPK4 were drilled subsequently in 1995 to create an HDR reservoir.

The GPK2 well was drilled to 5,000 m depth and stimulated in 2000. GPK3 was drilled in 2002 based on the seismic data obtained from the previous wells. The recorded bottom hole temperature of GPK3 was about 200°C. The distance between GPK2 and GPK3 was 650 m. Subsequently, in 2003 GPK3 was stimulated to create a reservoir between GPK2 and GPK3 (Baria et al., 2000, 2004). A circulation test was conducted between GPK2 and GPK3, with GPK2 as the production well. Initially the production rate declined in spite of the injection rate being kept constant. The loss of production was attributed to the continuous expansion of the reservoir. Nearly 30,000 to 50,000 m^3 were injected during the simulation period which resulted in mild seismic events. This suggests that the large injection was responsible for stress distribution and reflects the residual strain energy stored in the rocks. GPK4 was an inclined well

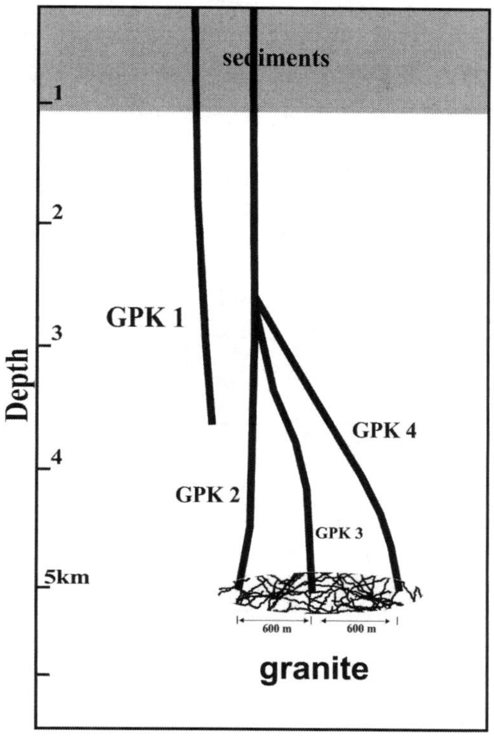

Figure 9.5 Schematic diagram showing the wells drilled in Soultz project (adapted from Baria et al., 2006).

drilled in 2003 to 5,260 m. During the circulation test using a tracer, the GPK2 well was able to yield 20% of the fluid injected, while GPK4 could yield only 2%. This shows that the connectivity between GPK3 and GPK2 was well established, while the connectivity between GPK3 and GPK4 was poor. The tracer recovered from GPK2 revealed the presence of a large volume of hydrothermal fluids present in the fractures in the granite at that depth. Although there were certain problems related to scaling, the Soultz EGS project was connected to the grid in 2011 and operated at 160°C/20 bars while the injection was at 70°C/18 bars, producing a gross 2.2 MWe using ORC technology. It is also planned that the project should supply 24 MWt to a starch factory (Scheiber et al., 2015, Vernier et al., 2015).

9.2.5 Cooper Basin, Australia

This Australian EGS initially started at the Hunter Valley, located north of Sydney due to large electricity market. This was started in 1999 as a joint venture between Pacific Power and the Australian National University in Canberra. The results and technology developed triggered interest in the geothermal project in Australia leading to the formation of a company, The Geodynamics Limited, floated on the stock exchange in

2002. Geodynamics acquired the Hunter Valley project later and shifted its interest to the Cooper Basin. However, under the current energy policy scenario of the government, the Hunter Valley project was abandoned. The focus was now on the Cooper Basin, where oil and gas drilling data indicated that the basement granite temperature at 4 km depth was about 250°C. The first well in the Cooper Basin was Habanero 1 (H01), an injection well, which was completed in 2003. Habanero is located near Innamincka NNE of Adelaide. The target reservoir here was a high heat generating granite lying below 3,600 m thick sediments. The well encountered granite 3,668 m that is under a high stress state for shear failure. Some of the fractures that H01 traversed were under very high pressure with water at 35 MPa. This over-pressure was countered by heavy drilling fluids. Since the fractures were highly permeable and it is also presumed that these rocks might have slipped along these fractures. Because of this permeability, nearly 250 m^3 of drilling fluids were lost (Hogarth et al., 2013). The second well, H02 drilled in 2004, encountered a problem due to a bridge plug that was lost in the hole. The third well H03 was drilled in 2008 successfully followed by H04 in 2012. Between 2008 and 2009, circulation was conducted between H01 and H03 which are 570 m apart. The temperature recorded at the surface was 212°C and the brine circulated was 61,000 tonnes with a flow rate of about 15 kg/s. In 2013 a circulation test between H01 and H04 was conducted by circulating 150,000 tonnes of brine at the rate of about 19 kg/s and the temperature recorded was 213°C (Hogarth et al., 2013).

The Cooper Basin project revealed several interesting findings: 1) a homogenous granite mass is an ideal candidate for an EGS project, 2) if the granite is hydrothermally altered, then the drilling becomes easy and perhaps may be reflected in the cost, 3) hydrothermal fluids can be encountered at those depths and open fractures/shear zones too occur at such depths which can get activated due to circulation and drilling, 4) an over thrust regime can be expected at such depths which helps the stimulation and creation of horizontal reservoirs. This will help to circulate fluids to increase the production rate and volume thus encouraging commercialisation of the technology (MIT, 2006).

9.2.6 Newberry Volcano EGS

Newberry Volcano is a shield volcano located in central, 65 km east of the Cascade Range.

This is a bi-model Quaternary volcano, active about 600,000 years ago. Its caldera diameter is about 8 km. The flows that erupted were, basaltic andesites, pyroclasts, silicic tuffs, and lahars. Silicic lavas and pyroclastics were erupted during the end phase (300,000 to 80,000 years). Exploration work for the EGS project in this site was started in 1970. The project site falls outside the Newberry National Park. The flanks of the volcano recorded a high geothermal gradient of 99°C/km while at deeper levels, exploration drill holes for geothermal gradients yielded 264°C/km. The first phase of the project commenced with the drilling of two wells, 55-29 and 46-16. Well 55-29 encountered multiple fractures with a bottom hole temperature of 315°C without any hydrothermal activity. Well 46-16 encountered open fractures filled with hydrothermal minerals. Well 55-29 encountered green schist facies metamorphosed volcanic rock at a depth of 1.9 km with a recorded bottom hole temperature of 315°C.

Well 46-16 is much deeper (3.5 km) and encountered green schist facies rocks over granodiorite, and the recorded bottom hole temperature was 315°C. The bore well also encountered dikes (Waibel et al., 2012).

9.3 EGS PROSPECTS IN THE COUNTRIES AROUND THE RED SEA

As discussed in Chapter 4, the geothermal provinces around the Red Sea evolved due to the breakup of the ANS and subsequent volcanic and tectonic activities related to the opening of the Red Sea. The remnants of the crystalline basement of the ANS, after the opening of the Red Sea, are found in Egypt (Figure 4.2), Eritrea (Figure 4.6), Yemen (Figure 4.23) and Saudi Arabia (Figure 4.3). A large part of the shield has been partitioned into the Arabian Shield covering an area of about 1 million km^2 (Nehlig et al., 2002). The granites and related rocks of the shield are characterised by a high uranium, thorium and potassium content. They, therefore, have the capacity to generate high heat. Due to the lack of data on the concentration of uranium and thorium in the granites occurring in Eritrea and Yemen, it is difficult to assess their heat generating capacity and their potential as EGS candidates. Next we evaluate the EGS potential of Egypt and Saudi Arabia.

9.3.1 EGS potential of Egypt

9.3.1.1 Radiogenic granites

Egypt, like Saudi Arabia, is endowed with post orogenic granites and its equivalents. The evolution of these granites were described in Chapter 4. These granites are found in the southern part of the Sinai as well as the south-eastern and south-western parts of the desert region (on the eastern and western side of the River Nile). In the south-eastern part of the region, the granites have intrusive contact with the basement crystallines (the Nubian shield high grade metamorphic rocks) and are intruded into the Neoproterozoic Arabian-Nubian Shield (ANS) crystallines. These granites are extremely rich in uranium and thorium and in certain cases host uranium mineralisation that is of economic importance (Saleh et al., 2014, Emam et al., 2011, Saleh, 2006, Raslan et al., 2012, Gaafar, 2014). These granites are enriched in accessory minerals like zircon and titanite which host the radioactive elements. For example, the concentration of uranium and thorium in these minerals occurring in granite intrusives from the Hafafit region (Figure 9.5) vary from 3,047 to 6,441 and 2,383 to 1,462 ppm respectively (Lundmark et al., 2012).

In the eastern desert, older and younger granites outcrop (Figure 9.6). The older granites, varying in age from 880 to 610 Ma include diorite to tonalite, trondhjemite and monzonite and are deformed. In Egypt the emplacement of these granites was related to Shaitian (850-800 Ma), Hafafit (760-710 Ma) and Meatiq (630 Ma) deformation phases (Lundmark et al., 2012, El Ramly and Akaad, 1960, Bentor, 1985, Hassan and Hashad, 1990). The younger granites are the post orogenic and syn-orogenic granites (600-530 Ma), and are alkaline to peralkaline in nature. These are plutons that are emplaced in the older crust along the strike-slip shear zones (Stern et al., 1984, Fritz et al., 1996, Breger et al., 2002, Moussa et al., 2008, Lundmark et al., 2012).

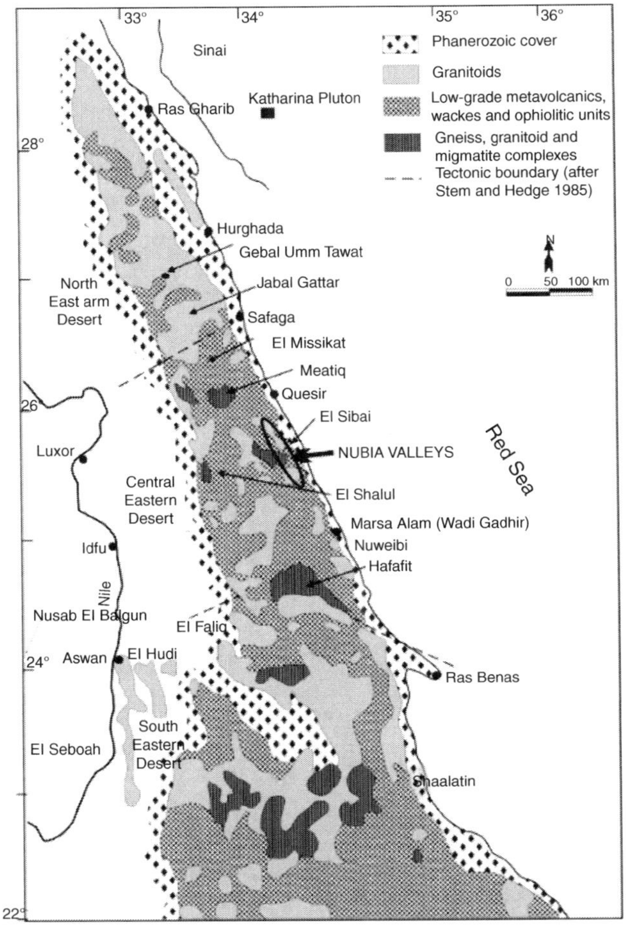

Figure 9.6 Map showing the distribution of granites and associated rocks in the eastern desert region (Lundmark et al., 2012).

A large number of granite intrusives occur in the south-western desert close to the Red Sea coast. The most interesting feature of these granites is that they are enriched in accessory minerals like zircon, titanite, monazite, thorite and uranothorite. These granitic melts were so enriched in uranium and thorium that these two elements entered the lattice of late stage crystallising minerals like fluorite. The uranium and thorium content in such fluorites (uranium rich fluorite) reported from the El Missikat mineralised granite (Figure 9.5) is about 2,200 and 7,500 ppm respectively (Raslan et al., 2009). In the same region, about 600 km from Aswan city, near Kukur and Dungul Oases, altered granites occur in association with sandstone of 400 Ma and 216 Ma old volcanics (Ibrahim et al., 2016). Small outcrops of the granites occur around the oases and the exposed area of the granites is about 3 km^2. They are covered by alluvium and Paleozoic sandstone and are intruded by basic dikes of Mesozoic age. These granites are highly altered and are silicified, kaolinised, hematitisation and sericitised. Because of the alteration, they are granular. The most characteristic feature of the silicified and kaolinised granites is that they are enriched in uranium and thorium (Ibrahim et al., 2016). The concentration of uranium and thorium in silicified granites varies from 5 to 954 ppm and 3 to 180 ppm respectively. Similarly, in the kaolinised granites, the uranium and thorium, vary from 38 to 96 ppm and 12 to 360 ppm respectively (Ibrahim et al., 2016). The heat generation by granites representing the entire south-eastern and south-western desert provinces and the heat flow values calculated for this region are shown in Table 9.3.

All these granite plutons, have roots spread over a long area, and their outcropping area is very small. On a regional scale they are covered by Post-Cambrian sedimentary rocks like sandstone, shale and chalk. For example, the stratigraphic sequence of area around Qusier (Figure 9.5), along the Red Sea coast, is shown in Figure 9.7.

It is apparent that the Precambrian basement with its intrusives is insulated by a >600 m thick sedimentary formation (Figure 9.6) which can contain the heat generated by the granites and associated rocks as well as the heat conducted by the mantle through conduction. A schematic subsurface cross section of the formations within the Nubia valleys (Figure 9.8), located SW of Qusier (Figure 9.5) is shown in Figure 9.9.

9.3.1.2 *Heat flow values*

Extensive heat flow and thermal gradient measurements have been made along the eastern margin in Egypt in the south-eastern desert extending up to the Red Sea axis (Evans and Tammemagi, 1974, Girdler, 1970, 1977, Gettings, 1982, Gettings and Showail, 1982, Morgan et al., 1977, 1981, 1983, 1985). The heat flow values measured over the sediments and granites and gneiss along the south-eastern desert are shown in Figure 9.9 and the heat flow profile over the land extending from the western margin of the Red Sea coast to a distance of about 200 km inside the south-western desert is shown in Figure 9.10. Because of the paucity of data on the uranium and thorium concentration in the granites from the south-eastern desert region, the reported heat flows were based on the then existing data and are therefore low (Morgan et al., 1985). For example, the El Hudi I type granites, located east of Aswan city (Figure 9.5), which generate 57 to 107 mW/m^2 (Table 9.3) of heat, were not considered by the earlier researchers while assessing the heat flow values of areas away from the Red Sea coast. The average heat generation by the granites is 18 $\mu W/m^3$ and the average heat flow value is 220 mW/m^2 (Table 9.3). This high value is apparently due to the occurrence of uranium rich mineral phases and secondary uranium deposits in the younger granites and related rocks in the south-eastern desert. Thus, the thermal anomaly of the entire south-eastern desert is higher by several folds than that reported by earlier researchers (Figures 9.10 and 9.11).

Thus, it is apparent that the granites and associated rocks occurring in Egypt in general and in the south-eastern region in particular are rich in heat generating radioactive elements and the heat generated by these granites are extremely high compared to normal granites. Granites of similar nature occur in the Arabian Shield, on the eastern side of the Red Sea.

Since all the landmasses around the Red Sea have a similar evolution history, with regards to tectonism and volcanism, granites, especially the post orogenic

Table 9.3 Uranium, thorium and potassium content and heat generated by the granites from Egypt. The heat flow is calculated assuming a 10 km thick granite slab. (Source: 1: Katzir et al., 2007, 2: Saleh et al., 2014, 3: Emam et al., 2011, 4. Saleh 2006, 5. Raslan et al., 2012, 6. Gaffar, 2014).

	Sample No	*U (ppm)*	*Th (ppm)*	*K (%)*	*RHP (uWm^{-3})*	*Heat Flow (mW/m^2)*	*Ref*
1	D163	5.0	27.0	4.2	3.5	75.4	1
2	D165	6.0	26.0	4.1	3.7	77.3	1
3	D82	6.0	23.0	4.2	3.5	75.2	1
4	D135	4.0	31.0	4.2	3.6	75.7	1
5	D97	10.0	40.0	3.9	5.7	97.0	1
6	D54	11.0	31.0	3.5	5.3	93.0	1
7	D59	9.0	30.0	3.7	4.7	87.3	1
8	D93	8.0	30.0	3.7	4.5	84.8	1
9	D43	9.0	36.0	3.6	5.1	91.4	1
10	D40	10.0	29.0	3.3	4.9	88.9	1
11	1.0	251.0	252.0	3.4	82.3	862.5	2
12	2.0	360.0	548.0	2.9	130.7	1346.9	2
13	3.0	230.0	414.0	3.1	88.0	920.2	2
14	4.0	373.0	571.0	2.6	135.6	1395.9	2
15	5.0	384.0	439.0	3.2	129.3	1333.5	2
16	6.0	240.0	432.0	3.6	91.9	958.9	2
17	10.0	2.0	5.0	3.0	1.1	51.4	3
18	11.0	8.0	13.0	3.1	3.2	72.4	3
19	12.0	7.0	17.0	2.8	3.2	72.4	3
20	17.0	9.0	15.0	3.2	3.7	76.5	3
21	19.0	13.0	24.0	3.5	5.3	93.3	3
22	3.0	17.0	31.0	2.7	6.8	107.7	3
23	6.0	14.0	33.0	2.8	6.1	101.4	3
24	40.0	6.0	22.0	3.4	3.4	73.8	3
25	41.0	9.0	30.0	2.9	4.7	86.6	3
26	43.0	13.0	29.0	3.2	5.6	96.4	3
27	46.0	7.0	23.0	3.5	3.7	77.2	3
28	48.0	8.0	27.0	3.0	4.2	82.0	3
29	23.0	5.0	17.0	2.7	2.7	67.2	3
30	26.0	7.0	18.0	2.9	3.3	73.2	3
31	32.0	4.0	13.0	2.7	2.2	61.8	3
32	33.0	6.0	16.0	2.8	2.9	69.1	3
33	34.0	2.0	7.0	1.9	1.2	51.7	3
34	36.0	4.0	13.0	2.6	2.2	61.7	3
35	38.0	5.0	15.0	3.1	2.6	66.1	3
36	39.0	3.0	10.0	2.9	1.7	57.4	3
37	1.0	6.0	22.0	4.3	3.5	74.7	4
38	2.0	4.0	17.0	4.2	2.6	66.0	4
39	3.0	12.0	30.0	3.4	5.5	94.7	4
40	4.0	13.0	34.0	4.1	6.1	100.8	4
41	5.0	10.0	27.0	4.0	4.8	88.2	4
42	6.0	10.0	36.0	4.6	5.5	94.9	4
43	7.0	14.0	50.0	4.9	7.5	115.1	4
44	8.0	34.0	22.0	4.6	10.7	146.9	4
45	9.0	25.0	26.0	3.7	8.6	125.7	4
46	10.0	14.0	20.0	4.1	5.4	93.7	4
47	11.0	26.0	22.0	4.0	8.6	125.8	4
48	1.0	7.0	13.3	3.4	3.0	70.3	5

(continued)

Table 9.3 Continued.

	Sample No	*U (ppm)*	*Th (ppm)*	*K (%)*	*RHP (uWm^{-3})*	*Heat Flow (mW/m^2)*	*Ref*
49	2.0	6.6	8.0	0.3	2.3	62.7	5
50	3.0	16.9	33.8	0.1	6.7	106.8	5
51	4.0	18.5	24.4	2.3	6.7	106.6	5
52	5.0	14.1	18.3	1.6	5.0	90.3	5
53	7.0	10.7	21.1	3.8	4.6	85.7	5
54	AN2	38.3	14.4	3.8	11.2	151.9	5
55	AN1	64.0	26.2	3.3	18.6	225.7	5
56	1S	84.0	12.0	3.0	22.7	267.1	5
57	2S	86.0	15.0	3.3	23.5	274.6	5
58	3S	85.0	8.0	3.6	22.7	267.3	5
59	4S	85.0	9.0	4.1	22.9	268.5	5
60	5S	78.0	13.0	3.2	21.3	252.5	5
61	11.0	80.0	7.0	2.8	21.3	253.1	5
62	12.0	90.0	15.0	2.8	24.4	284.4	5
63	13.0	83.0	5.0	2.6	21.9	259.2	5
64	14.0	78.0	7.0	3.1	20.8	248.3	5
65	15.0	85.0	9.0	3.0	22.8	267.5	5
66	A1	225.0	25.0	4.6	60.0	640.0	5
67	A2	210.0	28.0	5.1	56.4	603.9	5
68	Nwb1	5.4	18.5	3.7	3.0	70.1	6
69	Nwb2	5.4	18.2	3.4	3.0	69.7	6
70	Nwb3	6.5	19.3	4.2	3.4	74.0	6
71	Nwb4	5.8	23.4	5.1	3.6	75.9	6
72	Nwb5	5.1	16.8	3.6	2.8	68.1	6
73	Nwb6	3.7	23.2	4.1	2.9	69.4	6
74	Nwb7	4.7	26.7	3.9	3.4	74.2	6
75	Nwb8	2.8	23.1	3.9	2.7	66.8	6
76	Nwb9	4.8	26.6	4.0	3.4	74.5	6

granites, granites occurring in other countries like Eritrea and Yemen (where masses of ANS is represented) will have similar heat generating capacity and hence are worth investigation with respect to their suitability as an EGS source. Due to the paucity of data on the granites from other parts of the land masses around the Red Sea, a detailed interpretation and EGS assessment is not possible.

9.3.2 EGS potential of Saudi Arabia

The regional orogenic activities that resulted in the breakup of the Arabian-Nubian Shield (NAS), and the subsequent magmatic episodes before the initiation of the Red Sea rift, resulted in the evolution of high heat generating granites (see Chapter 4). The complex tectono-magmatic episodes that occurred over the ANS between 900 and 601 Ma and the rift-related volcanic activity between 30 and 5 Ma resulted in crustal thinning; under plating of the oceanic crust below the attenuated continental crust gave rise to high heat flow varying from 110 to 209 mW/m^2. Such high thermal anomalies due to the mantle conduction/convection and radiogenic heat generated by the granites gave rise to high temperature geothermal systems along the western margin of the Arabian Shield (Gettings et al., 1986, Lashin and Al Arifi, 2012, Chandrasekharam et al., 2015b).

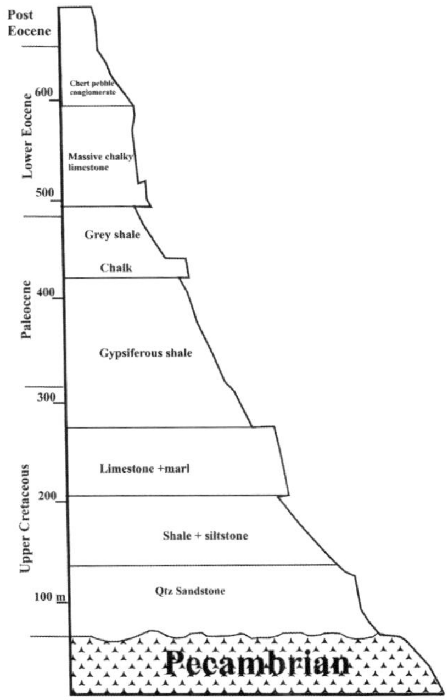

Figure 9.7 Stratigraphic column around Qusier showing the sedimentary formations above the Precambrian basement (adapted from Greene, 1984).

9.3.2.1 *Radiogenic granites*

As described in Chapter 4, the ANS comprises older crustal rocks and Neoproterozoic terranes caught in between the East and West Gondwana lands (Figure 4.27). These Neoproterozoic terranes are preserved in the eastern side of the Red Sea over the Arabian Shield (Figure 4.29). A large part of the ANS now represents the western Arabian Shield. This shield, as described, represents the "Wilson Cycle". Prior to the tectonic and magmatic events that reshaped the ANS between 900 and 400 Ma and later younger magmatic and tectonic events that accompanied the Red Sea rift, the ANS was considered as a juvenile crust evolved between East and West Gondwana (Figure 2.48), and representing amalgamated pieces of land and interwoven with several arcs and subduction tectonics remnants, represented by ophiolite belts that are exposed over the present day shield regions of Arabia and the western part of the Red Sea (Figure 4.29). Thus, the western Arabian Shield enclosed distinct accreted terranes, each terrane separated by ophiolite bearing suture zones (Figure 4.29) (Stoeser and Camp, 2014). These suture zones with ophiolites are still preserved and host several minerals and metals of economic value (Al-Shanti and Roobol, 1979, Ahmed and Habtoor, 2015).

The magmatic activities between 900 and 631 Ma resulted in a regional scale plutonism which is represented by rocks of granitic composition. The peak activity was between 660–610 Ma. The end stage of this activity is represented by peraluminous and peralkaline alkali feldspar granites, pegmatites and acid dikes. The entire plutonic

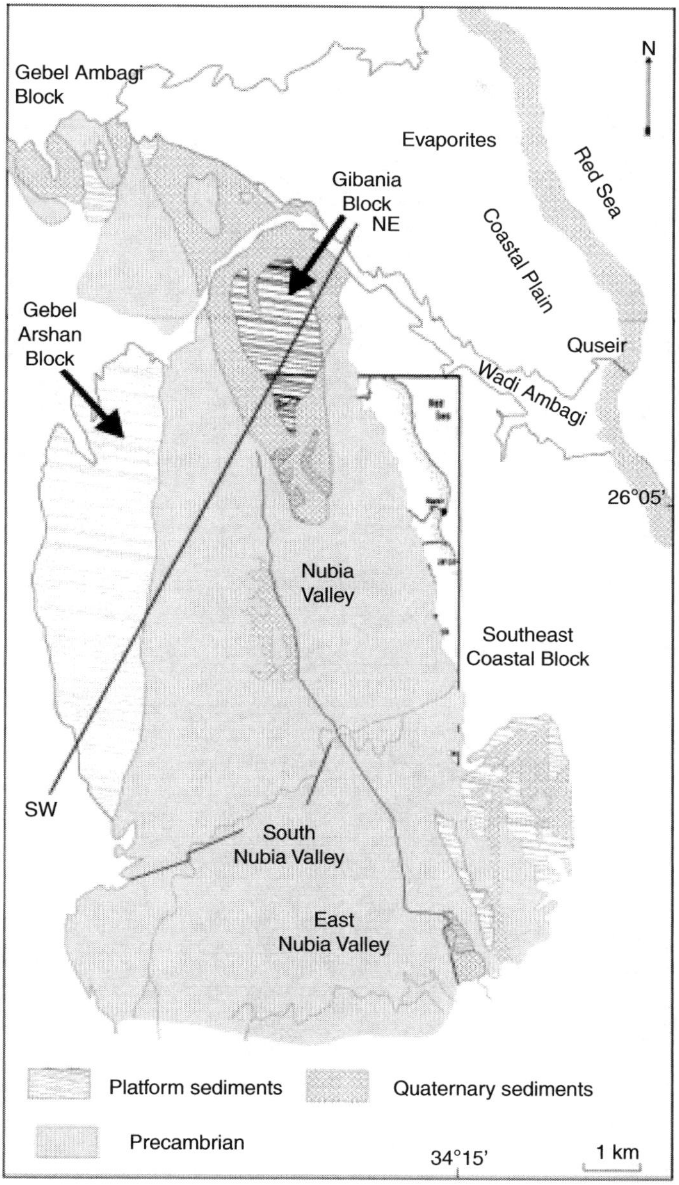

Figure 9.8 Geology of Nubia valleys. For the location of the Nubia valleys see Figure 9.5. A subsurface section along NW-SE line is shown in Figure 9.8).

activity is confined to the Arabian Shield which occupies an area of 770,000 km^2. The felsic plutonic rocks constitute 55% of this area and the rest is occupied by the volcanic rocks or the harrats. Out of these 55% of felsic plutonic rocks, granitic rocks represent 63% and the area occupied by the granitic rocks is 161,467 km^2 (Stoeser, 1986,

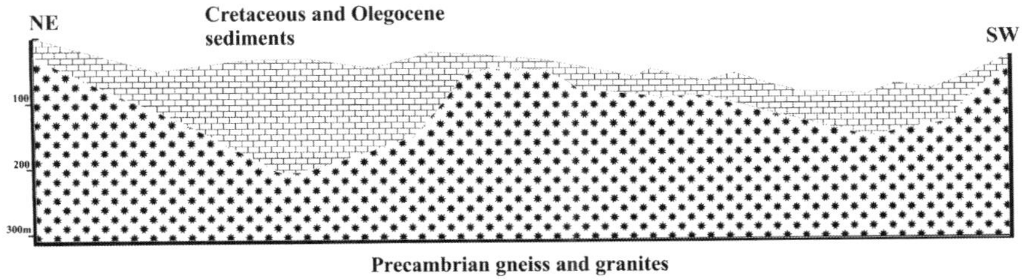

Figure 9.9 Schematic cross section across the Nubian valleys (line A-B, Figure 9.7). The Precambrian crystallines are intruded by younger granites and related rocks that are highly radiogenic (Table 9.3) (adapted from Greene, 1984).

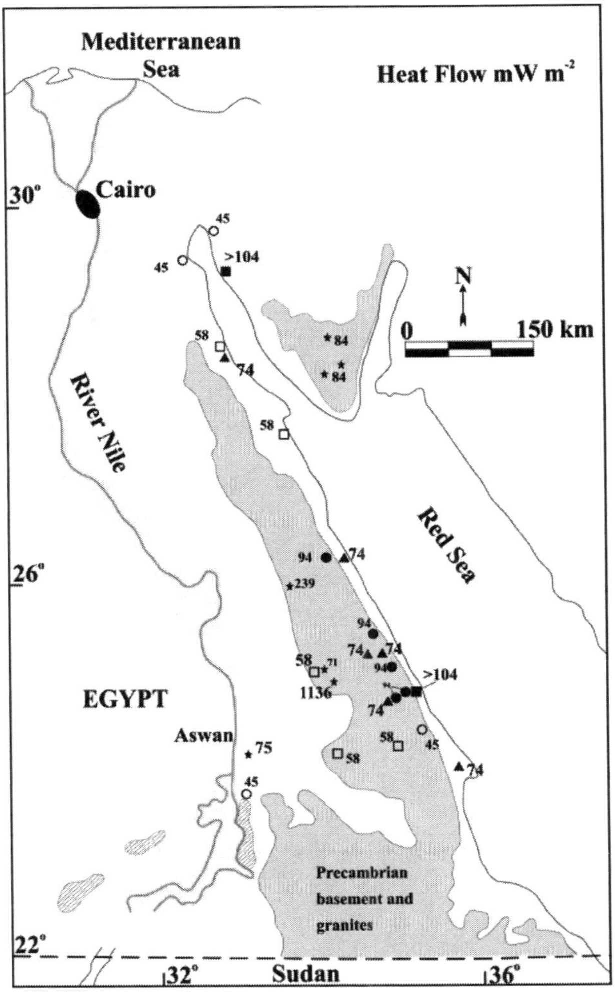

Figure 9.10 Heat flow values over the south-eastern desert region (adapted from Morgan et al., 1985).

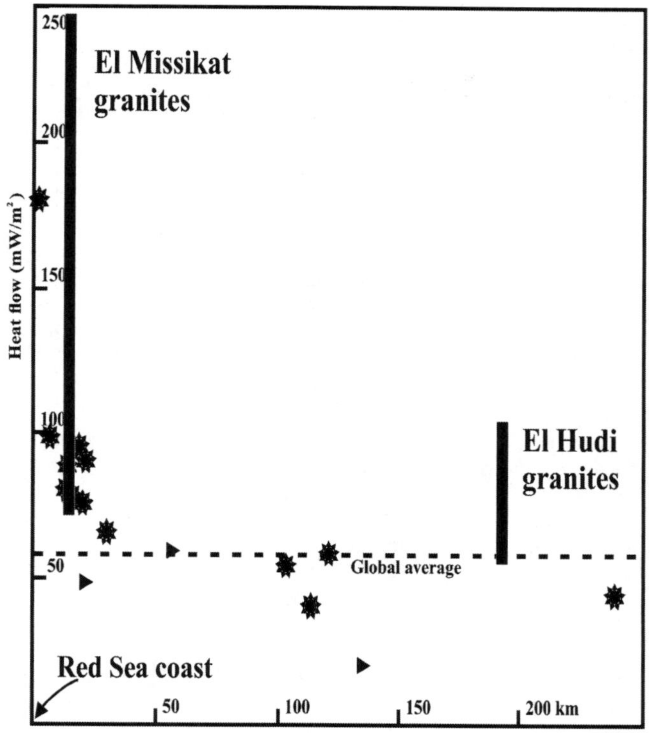

Figure 9.11 Heat flow profile across the south-eastern desert (south of 26° latitude, Figure 9.5), Egypt. Stars (granites) and triangles (sediments) were reported by Morgan et al. (1985). The heat flow values of El Hudi and El Missikat granites (Figure 9.5, Table 9.3) are included to show the wide heat flow values across the desert.

Chandrasekharam et al., 2014a,b, Chandrasekharam et al., 2015a,b,c,d). The distribution of granites and related rocks in the western shield was shown in Figure 4.33 (Chapter 4).

These paleo-suture zones are the loci of uranium-bearing minerals which were mobilised from the granite plutons shown in Table 4.7 (Stoeser, 1986, Stern and Johnson, 2010, Du Bray, 1986, Elliott, 1983, Bokhari et al., 1986, Agar, 1992, Dawood et al., 2010). These minerals in the granites and associated pegmatites gave rise to high temperature geothermal systems along the western Arabian Shield and maintain a high geothermal gradient and high heat flow value. These radioactive mineral deposits and the high concentration of radioactive elements make these granites very unusual. The heat flow regime is similar to the south-eastern desert region of Egypt discussed above. The uranium, thorium and potassium content in these high heat generating granites and the heat flow values calculated based on the concentration of these elements are shown in Table 9.4.

The locations of the high heat generating granites in the western Arabian Shield are shown in Figure 9.12.

Table 9.4 Uranium, thorium and potassium in the granites and related rocks in the western Arabian Shield and the heat generated by the granites and surface heat flow values over these rocks (adapted from Chandrasekharam et al., 2015d).

Reff	*Sample No.*	*Name of body*	*Lat (N)*	*Long (E)*	*U (ppm)*	*Th (ppm)*	*K (wt%)*	*RHP*	*HF (10 km thick body)*
2	184385	Al Bara bat ho l it h	23.48	41.34	4.1	14.0	3.9	2.38	123.83
	184412	Al Maslukah	24.47	44.23	3.5	18.0	3.8	2.49	124.87
	889494	Baid al Jimalah	25.15	42.69	13.3	35.2	3.9	6.22	162.15
	893039	Baid al Jimalah	25.15	42.69	11.8	24.4	3.7	5.07	150.71
	893159	Baid al Jimalah	25.15	42.69	11.3	29.6	4.7	5.39	153.90
	893235	Baid al Jimalah	25.15	42.69	10.3	28.9	4.0	5.02	150.17
	184122	Dukhnah	24.78	43.31	4.8	14.0	4.0	2.57	125.69
	184127	Dukhnah	24.93	43.16	6.6	20.3	3.9	3.46	134.55
	184241	Dukhnah	24.81	43.15	1.5	7.8	4.4	1.35	113.47
	184431	Dukhnah	24.85	43.16	7.1	13.2	4.2	3.11	131.14
	184432	Dukhnah	24.87	43.22	4.5	10.3	4.2	2.25	122.51
	184154	Gharnaq Pluton	23.62	40.55	5.5	12.5	3.7	2.62	126.19
	184154A	Gharnaq Pluton	23.62	40.55	5.5	11.5	3.8	2.57	125.66
	184155	Gharnaq Pluton	23.60	40.56	4.5	9.5	3.5	2.14	121.37
	184156	Gharnaq Pluton	23.55	40.57	3.5	11.5	3.4	2.03	120.27
	184382	Gharnaq Pluton	23.59	40.56	4.8	11.6	3.4	2.35	123.53
	184383	Jabal Ramram	23.46	40.69	2.3	5.6	2.6	1.23	112.28
	184004	Hadb ad Dayahin	23.54	41.17	2.8	2.7	3.5	1.24	112.37
	184171	Hadb ad Dayahin	23.59	41.19	13.7	28.5	2.4	5.72	157.16
	184180	Hadb ad Dayahin	23.57	41.23	5.5	4.1	3.3	2.01	120.05
	184184	Hadb ad Dayahin	23.52	41.23	5.2	4.2	3.1	1.92	119.17
	184381	Hadb ad Dayahin	23.51	41.22	2.8	8.2	3.9	1.65	116.46
	184415	Hadbat Tayma	24.14	41.23	8.5	26.7	4.2	4.41	144.14
	184384	Hudayb al Jidar	23.37	41.18	5.5	12.2	3.3	2.57	125.70
	184014	Jabal Abha	24.41	40.69	6.9	20.8	3.3	3.52	135.19
	184015	Jabal Abha	24.40	40.69	8.5	22.1	2.8	3.97	139.66
	184018	Jabal Abha	24.39	40.77	9.1	29.8	3.4	4.72	147.23
	184374	Jabal Abha	24.36	40.68	0.5	2.0	0.5	0.33	103.28
	184375	Jabal Abha	24.38	40.67	7.7	25.6	3.3	4.06	140.63
	184019	Jabal Aj i r	24.98	41.12	2.8	9.6	3.9	1.76	117.58
	184021	Jabal Aj i r	24.97	41.14	2.6	7.7	4.1	1.58	115.79
	184023	Jabal Aj i r	24.86	41.17	3.7	9.1	4.0	1.94	119.44
	184089	Jabal Dhuray	24.22	43.72	15.8	35.2	3.7	6.85	168.46
	184414	Jabal Dhuray	24.22	43.72	4.4	13.0	3.0	2.31	123.08
	184031	Jabal Furqayn (north)	24.82	41.68	3.3	6.9	3.9	1.68	116.84
	184036	Jabal Furqayn (north)	24.90	41.60	2.1	6.5	3.7	1.35	113.46
	184038	Jabal Furqayn (north)	24.93	41.68	2.1	7.5	3.9	1.43	114.30
	184377	Jabal Furqayn (north)	24.88	41.54	2.3	6.1	3.9	1.38	113.80
	184378	Jabal Furqayn (north)	24.91	41.66	2.6	7.4	4.0	1.56	115.57
	184040	Jabal Furqayn (south)	24.06	41.25	7.6	21.3	3.8	3.79	137.92
	184379	Jabal Habhab al Jissu	24.52	41.61	8.0	15.0	3.8	3.45	134.53
	184380	Jabal Habhab al Jissu	24.46	41.59	12.7	28.5	3.7	5.58	155.81
	184009	Jabal Hadb ash Sharar	23.81	40.97	8.1	21.9	3.2	3.90	138.96
	184056	Jabal Hadb ash Sharar	23.77	40.98	7.2	12.3	3.1	2.99	129.86
	184396	Jabal Hadb ash Sharar	23.79	40.99	0.7	2.0	1.0	0.41	104.06
	184087	Jabal Jabalah	24.78	43.89	4.2	10.1	4.0	2.15	121.49
	184088	Jabal Jabalah	24.80	43.89	6.8	15.6	3.9	3.19	131.94
	184269	Jabal Jabalah	24.78	43.86	8.0	8.5	4.1	3.02	130.19
	184070	Jabal Haslah	23.82	42.05	2.4	7.6	3.9	1.51	115.11
	184177	Jabal Haslah	23.83	42.01	2.6	8.0	4.0	1.58	115.84

(*Continued*)

Table 9.4 Continued.

Reff	Sample No.	Name of body	Lat (N)	Long (E)	U (ppm)	Th (ppm)	K (wt%)	RHP	HF (10 km thick body)
	184388	Jabal Haslah	23.86	42.00	4.4	15.8	4.0	2.59	125.89
	184389	Jabal Haslah	23.83	42.00	3.0	9.3	3.8	1.77	117.69
	184390	Jabal Haslah	23.83	42.00	3.3	11.5	3.5	1.97	119.68
	184329	Jabal Khazaz	25.42	43.57	6.3	27.5	4.3	3.91	139.15
	184334	Jabal Khazaz	25.38	43.55	3.9	28.6	3.9	3.34	133.37
	114337	Jabal Khazaz	25.32	43.60	6.2	21.9	3.5	3.44	134.44
	184338	Jabal Khazaz	25.33	43.60	9.1	25.5	3.6	4.44	144.36
	184433	Jabal Khazaz	25.42	43.57	3.8	17.4	4.2	2.58	125.85
	184434	Jabal Khazaz	25.38	43.58	3.2	18.7	4.2	2.52	125.17
	184435	Jabal Khazaz	25.33	43.60	7.4	13.9	3.6	3.21	132.05
	184419	Jabal Khurs	23.58	44.69	7.6	19.7	3.9	3.68	136.76
	184077	Jabal Khinzir	23.17	43.82	30.3	8.3	3.5	8.69	186.90
	184079	Jabal Khinzir	23.15	43.84	20.0	22.8	3.5	7.05	170.47
	184080	Jabal Khinzir	23.14	43.87	9.8	27.1	3.6	4.74	147.36
	184129	Jabal Khinzir	23.16	43.86	10.3	21.6	4.3	4.54	145.43
	184130	Jabal Khinzir	23.15	43.85	10.6	24.2	4.0	4.78	147.76
	184420	Jabal Khinzir	23.16	43.79	12.2	12.3	3.4	4.31	143.06
	184421	Jabal Khinzir	23.15	43.81	12.2	26.8	4.0	5.36	153.62
	184422	Jabal Khinzir	23.13	43.90	2.3	4.8	2.7	1.17	111.69
	184117	Jabal Minya	24.99	43.40	16.1	24.1	4.4	6.22	162.16
	184359	Jabal Minya	24.99	43.39	14.2	37.4	3.8	6.59	165.93
	184373	Jabal Minya	24.99	43.39	23.7	38.5	3.7	9.10	190.97
	184423	Jabal Minya	24.97	43.35	11.2	27.2	4.1	5.14	151.41
	184424	Jabal Minya	24.97	43.35	13.7	27.1	4.2	5.79	157.90
	184425	Jabal Minya	24.96	43.33	11.6	15.9	2.9	4.35	143.54
	184376	Jabal Sanaa	24.73	41.29	3.3	8.3	3.8	1.78	117.82
	184439	Jabal Shiib	24.08	42.53	7.1	42.1	3.9	5.10	150.97
	184440	Jabal Shiib	24.16	42.44	13.8	47.1	3.9	7.17	171.73
	184440A	Jabal Shiib	24.16	42.44	14.6	44.0	4.0	7.17	171.69
	184441	Jabal Shiib	24.08	42.42	7.6	17.0	4.2	3.53	135.26
	184061	Jabal Saqrah	23.08	42.85	1.8	20.1	5.4	2.35	123.51
	184063	Jabal Saqrah	23.06	42.87	1.8	8.6	5.7	1.58	115.83
	184064	Jabal Saqrah	23.02	42.97	1.4	7.7	4.4	1.31	113.12
	184065	Jabal Saqrah	23.06	43.03	2.6	13.5	4.6	2.04	120.43
	184301	Jabal Saqrah	23.10	42.97	3.6	12.4	4.1	2.16	121.56
	184302	Jabal Saqrah	23.08	42.98	1.7	6.5	4.1	1.27	112.70
	184402	Jabal Saqrah	23.07	42.82	1.8	24.0	5.5	2.65	126.48
	184403	Jabal Saqrah	23.10	42.94	5.6	32.0	4.5	4.06	140.62
	184404	Jabal Saqrah	23.10	42.94	26.3	49.8	4.1	10.58	205.83
	184405	Jabal Saqrah	23.06	43.00	3.6	20.7	3.9	2.71	127.12
	184406	Jabal Saqrah	23.03	43.02	4.0	14.1	3.5	2.32	123.19
	181859	Jabal Si Isl lah	26.09	42.65	12.1	36.1	3.6	5.94	159.43
	181871	Jabal Si Isl lah	26.09	42.66	8.4	33.9	3.5	4.83	148.32
	181872	Jabal Si Isl lah	26.08	42.66	13.8	22.3	2.9	5.36	153.57
	181964	Jabal Si Isl lah	26.10	42.65	6.3	21.7	4.0	3.49	134.90
	184392	Jabat Tukhfah	23.88	42.16	9.9	25.0	3.7	4.61	146.05
	184393	Jabat Tukhfah	23.94	42.12	7.5	25.2	4.1	4.05	140.53
	184394	Jabat Tukhfah	23.94	42.13	7.9	25.4	4.0	4.17	141.67
	184391	Jabal Umm Adban	23.84	42.13	2.2	7.1	4.0	1.42	114.17
	184417	Jabal Za'abah	23.81	44.77	3.9	11.5	3.7	2.13	121.28
	184418	Jabal Za'abah	23.77	44.77	5.6	17.5	4.3	3.05	130.53
	184444	Jabal al Mudayh	24.41	42.21	8.2	38.4	4.2	5.16	151.61

(Continued)

Table 9.4 Continued.

Reff	Sample No.	Name of body	Lat (N)	Long (E)	U (ppm)	Th (ppm)	K (wt%)	RHP	HF (10 km thick body)
	184445	Jabal al Mudayh	24.45	42.20	7.6	24.9	4.1	4.05	140.46
	184397	Jabal al Yanufi	23.43	43.07	4.6	13.1	3.7	2.43	124.30
	184398	Jabal al Yanufi	23.45	43.07	2.8	14.3	3.7	2.07	120.65
	184399	Jabal al Yanufi	23.41	43.08	2.8	15.4	3.9	2.15	121.48
	184058	Jabal at Safuah	23.14	42.29	3.3	13.1	3.8	2.10	121.00
	184059	Jabal at Safuah	23.13	42.29	5.3	15.8	4.1	2.84	128.43
	184395	Jabal at Safuah	23.12	42.27	8.1	20.2	3.8	3.83	138.34
	184339	Ha lik Granite	25.17	43.65	1.6	25.3	4.3	2.57	125.72
	184342	Ha lik Granite	25.13	43.79	3.3	18.7	4.8	2.59	125.92
	184344	Ha lik Granite	25.18	43.74	3.8	28.9	4.1	3.36	133.64
	184149	Hiskah 1	24.62	42.94	5.9	18.9	4.1	3.20	132.01
	184426	Hiskah 1	24.68	42.85	7.6	24.4	4.4	4.05	140.53
	184242	Hiskah 1	24.59	43.08	9.5	29.0	4.3	4.85	148.48
	184368	Hiskah 2	24.82	42.90	0.4	14.6	5.9	1.68	116.78
	184427	Hiskah 2	24.91	42.80	7.2	16.5	4.1	3.38	133.75
	184446	Hiskah 2	24.93	42.57	12.4	25.8	3.8	5.33	153.29
	184447	Hiskah 2	24.93	42.59	11.8	19.5	4.1	4.77	147.69
	184152	Hiskan 3	24.59	43.17	3.0	9.2	4.0	1.79	117.87
	184428	Hiskan 3	24.59	43.17	2.6	12.9	4.5	1.98	119.75
	184429	Hiskan 3	24.58	43.22	2.0	8.4	4.3	1.51	115.09
	184430	Hiskan 3	24.59	43.17	2.4	8.3	4.2	1.59	115.90
	184409	Najirah	24.23	44.43	4.6	18.9	4.4	2.90	129.04
	184410	Najirah	24.23	44.44	5.1	16.1	4.3	2.82	128.16
	184411	Najirah	24.26	44.44	4.6	17.2	4.3	2.76	127.63
	184436	Suva j	25.20	43.29	11.9	24.2	3.9	5.10	150.98
	184437	Suva j	25.20	43.29	9.8	19.2	4.4	4.26	142.59
	184438	Suva j	25.18	43.28	7.9	21.2	4.3	3.89	138.90
	184274A	Unnamed 1	24.15	43.88	3.7	19.3	4.1	2.66	126.59
	184275	Unnamed 1	24.20	43.92	3.6	15.6	4.2	2.39	123.88
	184413	Unnamed 1	24.16	43.93	8.3	19.9	4.2	3.91	139.12
	184400	Unnamed 2	23.46	42.68	3.0	17.9	4.2	2.40	123.95
	184442	Unnamed 3	24.35	42.43	1.7	7.7	6.4	1.57	115.75
	184443	Unnamed 3	24.35	42.44	12.4	19.3	3.8	4.88	148.82
3	G-1	Salmon			4.3	10.3	4.8	2.26	122.58
	G-2	Bianco sardo			7.7	12.1	4.5	3.23	132.25
	G-3	Sweet gold			0.2	0.1	0.0	0.05	100.53
	G-4	Najran brown			0.0	0.0	0.0	0.00	100.05
	G-5	Saudi brown			0.0	0.1	0.0	0.01	100.08
	G-6	Saudi green			0.0	0.0	0.0	0.00	100.02
	G-7	Royal gold			0.0	0.0	0.0	0.00	100.01
	G-8	Spring green			0.0	0.0	0.0	0.00	100.02
6	1	Midian Granite	27.63	36.31	5.0	11.0	4.1	2.43	124.32
	2	Midian Granite	27.75	36.17	5.0	1.0	3.8	1.71	117.15
	3	Midian Granite	27.89	36.17	1.0	2.0	3.5	0.73	107.27
	4	Ghurayyah Granite	28.27	36.05	104.0	625.0	2.6	70.18	901.76
	5		28.26	36.17	88.0	160.0	2.6	33.93	539.26
	6	Vain	29.00	35.58	363.0	590.0	1.7	134.24	1542.42
7	7	Precursor granites	17.93	44.00	8.0	30.0	4.2	4.53	145.28
	8	Precursor granites	22.19	44.81	6.0	30.0	4.1	4.00	139.98
	11	Tin-bearing granites	18.09	44.01	12.0	48.0	3.2	6.71	167.06
	14	Tin-bearing granites	21.23	44.04	4.0	26.0	3.6	3.16	281.62
	15	Tin-bearing granites	22.10	44.71	12.0	34.0	3.4	5.75	307.50

(Continued)

Table 9.4 Continued.

Reff	Sample No.	Name of body	Lat (N)	Long (E)	U (ppm)	Th (ppm)	K (wt%)	RHP	HF (10 km thick body)
8	155575	J Mis	20.05	41.22	3.5	13.5	3.3	2.13	271.30
	155576		20.06	41.22	3.5	9.3	3.9	1.92	269.18
	155577		20.05	41.22	4.3	14.1	3.6	2.43	274.30
	155578	J. Shada	19.84	41.31	3.5	9.1	3.6	1.87	268.67
	155579		19.84	41.31	3.8	7.0	3.7	1.79	267.92
	155587		19.85	41.32	1.9	10.1	2.9	1.46	264.58
	155588		19.85	41.32	1.4	6.5	2.0	1.01	260.10
	155589		19.86	41.33	3.6	12.0	3.9	2.12	271.25
	155580	J. Durraa	18.77	42.27	4.9	12.8	4.2	2.55	275.49
	155581		18.77	42.27	5.1	15.3	4.6	2.81	278.06
	155582		18.78	42.30	3.7	12.7	4.4	2.25	272.54
	155583	J. Barquq	18.80	42.19	1.1	6.1	4.2	1.09	260.88
	155584		18.80	42.20	6.4	6.6	3.8	2.47	274.71
	155585		18.80	42.20	5.6	13.4	4.1	2.76	277.56
	155586		18.79	42.20	1.4	1.8	3.8	0.83	258.26
	155591	Al Ajarda	19.17	41.97	0.8	3.3	4.3	0.83	258.34
	155592		19.17	41.97	1.0	3.1	3.6	0.80	257.99
	155593		19.12	41.96	0.1	0.4	4.6	0.47	254.71
	155594		19.12	41.96	0.1	0.5	4.4	0.46	254.65
	155595		19.11	42.00	0.3	0.6	2.1	0.31	253.06
	155596	J. Fuqa'ah	19.05	41.44	4.0	7.1	3.8	1.87	268.73
	155597	Par al Jabal	19.84	41.54	1.0	2.2	1.6	0.55	255.49
	155598		19.84	41.55	1.0	0.8	1.5	0.46	254.62
	1555981		19.84	41.55	0.3	0.8	1.7	0.29	252.94
	155599	J. Duqayna	19.72	41.63	10.2	2.0	3.8	3.12	281.16
	155600		19.71	41.67	6.1	13.4	3.9	2.86	278.59
	155601	J. Raft	20.46	41.97	4.0	7.3	3.4	1.85	268.50
	155502		20.49	42.01	4.4	5.4	2.8	1.78	267.76
	155603		20.51	42.02	3.7	4.3	2.8	1.51	265.09
	155604		20.50	41.93	5.4	8.8	3.3	2.31	273.09
	155605		20.48	41.93	5.2	5.2	3.2	1.99	269.88
	155536		20.44	42.01	2.6	2.7	0.7	0.93	259.31
	155607	Al Mu'taridah	20.31	42.21	11.7	2.2	2.2	3.36	283.62
	155608		20.31	42.21	1.3	0.9	0.9	0.47	254.73
	155609		20.24	42.24	8.9	12.1	3.7	3.47	284.69
	155510		20.22	42.19	8.1	13.5	3.8	3.36	283.64
	155611		20.21	42.21	14.0	19.8	3.5	5.30	302.99
	155612	J. Balas	19.81	41.87	10.1	17.8	4.0	4.20	292.01
	155613		19.81	41.89	6.4	18.0	4.1	3.26	282.60
	155514		19.86	41.89	7.1	18.0	4.0	3.44	284.35
	155615	Wadi Shumms	19.90	41.86	6.9	19.2	4.3	3.51	285.10
	155516		19.89	41.92	2.4	8.8	4.0	1.61	266.07
	155517		19.95	41.88	1.2	4.2	4.3	0.99	259.87
	155518	J. Abu Sadi	20.42	40.01	3.7	12.8	3.9	2.20	272.02
	155619		20.43	40.01	5.7	10.5	4.1	2.58	275.82
	155620		20.45	40.02	2.9	11.8	3.7	1.91	269.08
	155621		20.41	40.04	3.9	14.0	4.0	2.34	273.36
	155622	Adan Pluton	20.39	40.87	1.3	6.7	2.3	1.02	260.19
	155523		20.39	40.87	1.4	6.7	3.8	1.18	261.82
	155624		20.39	40.87	0.8	5.6	2.7	0.84	258.43
	155625		20.40	40.93	2.3	7.6	3.5	1.45	264.51
	155626		20.40	40.93	1.8	7.0	3.8	1.30	263.02

(*Continued*)

Table 9.4 Continued.

Reff	Sample No.	Name of body	Lat (N)	Long (E)	U (ppm)	Th (ppm)	K (wt%)	RHP	HF (10 km thick body)
	155627	J. Ibrahan	20.44	41.17	1.4	3.6	3.6	0.96	259.55
	155626		20.44	41.17	1.1	4.4	3.4	0.91	259.14
	155629		20.43	41.14	1.2	6.9	3.6	1.13	261.27
	155530		20.42	41.13	1.2	4.5	3.4	0.94	259.39
	155631		20.41	41.15	2.0	6.0	3.1	1.21	262.12
	155632	Hawil	20.96	41.36	2.2	9.3	4.2	1.61	266.11
	155633		20.93	41.36	1.5	4.9	3.5	1.04	260.43
	155634		20.91	41.35	1.4	4.5	6.0	1.23	262.33
	155635		20.86	41.29	1.4	4.3	4.2	1.06	260.59
	155641		20.88	41.33	1.4	5.4	4.6	1.17	261.66
	155636	J An	21.29	41.17	2.2	9.3	4.3	1.62	266.22
	155537		21.29	41.18	1.5	4.9	4.1	1.10	261.03
	155638		21.29	41.18	1.4	4.5	4.1	1.06	260.62
	155639		21.29	41.18	1.4	4.0	4.2	1.04	260.39
	155640		21.29	41.18	1.4	5.4	4.2	1.13	261.32
	155542	J. Ounnah	21.07	41.15	6.2	19.6	3.5	3.29	282.86
	155643		21.05	41.12	0.1	0.1	0.2	0.04	250.37
	155644		21.05	41.12	0.1	0.1	0.2	0.04	250.38
	155645	Al Mahdan	20.94	40.88	2.0	8.5	2.8	1.35	263.52
	155646		20.89	40.88	1.3	7.3	4.1	1.21	262.07
	155647		20.90	40.86	0.7	0.8	4.8	0.68	256.78
	155648	J. Bargatinah	20.80	39.86	1.5	5.6	3.4	1.10	260.97
	155649		20.80	39.86	2.2	6.4	3.2	1.30	263.02
	155650		20.80	39.86	2.5	6.4	3.6	1.43	264.27
	155651		20.79	39.85	1.7	4.1	2.6	0.96	259.60
	155652	Unnaaed	21.16	39.96	2.8	6.8	4.5	1.61	266.09
	155653		21.15	39.95	2.6	9.6	3.5	1.66	266.55
	155654		21.16	39.97	3.5	5.5	3.6	1.63	266.26
	155655		21.18	39.98	2.0	5.1	3.2	1.17	261.66
	155656		21.18	39.98	3.3	6.7	4.0	1.68	266.81
	155657		21.18	39.98	2.9	8.5	3.8	1.70	266.96
	155558	Judayrah	21.39	40.42	5.1	55.2	4.5	5.54	305.40
	155659		21.45	40.43	2.8	13.3	4.7	2.08	270.84
	155660	J. Sha'ir	21.32	40.30	5.0	12.2	4.0	2.50	274.99
	155661	Unnaned	21.17	40.10	1.1	6.2	2.9	0.98	259.77
	155662		21.17	40.15	3.7	7.0	3.4	1.76	267.61
	155663	J. Abu Sibal	21.33	39.63	1.3	3.9	2.6	0.86	258.61
	155664		21.33	39.63	1.9	3.7	2.3	0.95	259.46
	155665		21.33	39.63	1.6	3.6	2.2	0.87	258.68
	155666		21.33	39.63	1.8	4.0	2.4	0.95	259.46
	155667	Unnaned	21.09	40.00	2.5	7.6	4.1	1.56	265.57
	155668		21.09	40.00	2.7	10.7	3.4	1.74	267.39
	155669	Unnaned	21.09	45.10	1.7	11.7	3.9	1.61	266.12
	155670	J. Savdah	20.80	40.22	1.7	12.1	4.5	1.70	266.99
	155571		20.80	40.22	1.2	3.0	3.9	0.89	258.90
	155672		20.79	40.24	1.5	10.0	4.3	1.48	264.77
	155673		20.79	40.24	2.0	9.3	2.7	1.40	264.01
	155674	J. Alonsa	20.49	40.67	0.6	1.6	1.8	0.42	254.23
	155675		20.49	40.67	0.5	1.7	1.7	0.39	253.92
	155676	Unnmmed	20.54	40.96	3.6	9.5	4.0	1.95	269.53
	155677		20.54	40.96	3.2	15.7	4.3	2.30	273.02

(Continued)

Table 9.4 Continued.

Reff	Sample No.	Name of body	Lat (N)	Long (E)	U (ppm)	Th (ppm)	K (wt%)	RHP	HF (10 km thick body)
	155578	J. Majiah	20.27	42.70	3.8	12.7	4.1	2.24	272.39
	155679		20.28	42.65	2.3	11.6	5.4	1.90	269.05
	155660		20.28	42.62	2.3	7.2	4.4	1.51	265.10
	155661		20.28	42.62	2.1	7.3	5.1	1.51	265.13
	155682	Sha'ib Dahthami	20.42	41.41	2.7	7.0	3.1	1.45	264.53
	155683		20.43	41.37	1.7	3.8	3.6	1.05	260.48
	155684	J. Suliy	20.47	41.52	6.8	15.9	4.6	3.27	282.73
	155665		20.98	42.50	6.4	16.1	4.6	3.19	281.85
	155686		20.98	42.50	5.3	17.2	4.7	2.99	279.86
	155687		20.97	42.53	3.0	7.7	4.1	1.70	266.97
	155588	J. Kor	20.99	42.75	5.1	8.6	2.9	2.19	271.86
	155669		20.98	42.81	6.5	13.4	4.3	3.00	280.02
	155690		20.95	42.85	6.1	20.2	4.3	3.37	283.75
	155691		20.98	42.85	6.3	10.8	4.1	2.75	277.55
	155692	J. Taweel	20.90	42.91	1.0	2.4	3.9	0.78	257.82
	155693		20.84	42.92	1.8	5.7	4.2	1.25	262.48
	155694		20.83	42.95	2.6	5.5	4.1	1.44	264.37
	155695		20.78	42.96	2.9	4.3	4.0	1.41	264.12
	155696		20.73	42.94	1.9	3.8	3.9	1.13	261.30
	155697	J. Bafdeh	19.56	42.95	18.3	21.1	3.4	6.49	314.85
	155698		19.55	42.93	4.6	14.5	3.7	2.53	275.30
	155699		19.54	42.89	3.3	9.8	3.7	1.86	268.61
	155700	Unnaaed	19.30	42.90	0.8	0.6	1.6	0.40	254.01
	155701	Jazirah	19.05	42.91	1.6	4.0	4.2	1.09	260.95
	155702		19.02	42.92	0.5	0.5	4.8	0.61	256.12
	155703		19.02	42.95	0.4	0.5	0.6	0.19	251.90
	155704		19.04	42.96	0.4	0.5	5.6	0.67	256.75
	155705	J. Marub	19.11	42.60	2.8	10.2	3.0	1.71	267.08
	155706		19.13	42.61	2.3	8.9	3.4	1.53	265.27
	155707	Unnaned	19.33	42.62	17.2	2.4	3.2	4.89	298.86
	155708		19.35	42.58	5.7	16.9	4.0	3.01	280.10
	155709		19.35	42.58	7.0	15.5	3.8	3.23	282.28
	155710		19.36	42.62	25.9	3.1	3.0	7.15	321.53
	155711		19.36	42.62	23.6	3.2	3.3	6.59	315.94
	155712	Madmbiyah Pluton	19.41	42.62	2.8	8.3	4.2	1.68	266.77
	155713		19.49	42.66	5.9	19.3	3.9	3.23	282.29
	155711	Gslalah Dome	19.56	42.67	0.9	5.5	4.0	0.98	259.78
	155715		19.57	42.68	1.4	3.3	4.0	0.97	259.71
	155716		19.63	42.70	6.2	19.1	3.6	3.24	282.42
	155717		19.63	42.70	15.1	16.2	2.1	5.20	301.96
	155718	Al Hideb	19.69	42.80	3.7	10.3	4.2	2.06	270.55
	155719		19.72	42.80	6.9	15.4	4.1	3.21	282.14
	155720		19.72	42.80	6.0	11.8	4.1	2.73	277.30
	155721	J. Munireh	20.60	42.78	2.3	4.9	4.3	1.33	263.28
	155722		20.60	42.78	1.4	3.5	4.5	1.03	260.25
	155724		20.56	42.71	3.7	9.3	4.1	1.99	269.91
	155724		20.51	42.71	6.8	8.6	4.0	2.70	277.02
	155725	J. al Jafar	20.51	42.79	7.7	2.2	4.2	2.52	275.21
	155726		20.51	42.78	7.3	7.8	4.1	2.79	277.94
	155726A		20.66	42.78	8.7	2.4	4.2	2.80	277.99
	155727		20.46	42.78	4.2	13.3	4.3	2.41	274.08
	155728	J. al Fu'ad	20.99	43.79	6.0	16.5	4.2	3.08	280.77
	155729		20.99	43.79	5.7	23.6	3.8	3.46	284.59

(Continued)

Table 9.5 Continued.

Reff	Sample No.	Name of body	Lat (N)	Long (E)	U (ppm)	Th (ppm)	K (wt%)	RHP	HF (10 km thick body)
	155730		20.95	43.83	5.1	14.1	4.3	2.69	276.90
	155731		20.93	43.85	4.6	15.3	4.1	2.63	276.30
	155732	J. al Qarah	20.27	43.21	1.6	7.6	3.6	1.29	262.88
	155733		20.27	43.21	4.5	8.6	3.9	2.13	271.32
	155734		20.22	43.23	4.9	9.8	4.0	2.31	273.10
	155735	Bnni Shawhatah	20.10	43.45	3.5	10.4	4.0	1.98	269.80
	155736		20.04	43.46	3.1	10.4	4.1	1.91	269.14
	155737	Unnaned	19.73	42.58	5.8	15.0	4.0	2.91	279.05
	155738	Wadi al Khanaq	19.76	42.77	12.4	26.4	3.9	5.37	303.75
	155739	Sirat Rishah	19.88	42.70	3.4	13.8	3.4	2.16	271.55
	155742	J. Asbah	20.15	41.95	0.7	4.5	5.0	0.95	259.54
	155743		20.10	41.92	2.2	5.1	4.1	1.29	262.91
	155744	J. Amoudah	20.08	42.80	4.8	10.6	4.1	2.34	273.43
	155745		20.13	42.79	3.8	9.6	4.2	2.04	270.43
	155746		20.22	42.79	1.7	3.9	4.2	1.10	260.96
	155747	J. Khashaadheed	20.25	42.88	4.3	11.5	4.1	2.29	272.90
	155748		20.32	42.89	13.0	20.0	4.0	5.10	301.03
	155749		20.32	42.89	6.9	15.7	4.1	3.23	282.34
	155750	J Umm Hashiyah	20.02	42.92	3.6	13.3	3.1	2.14	271.42
	155751		19.98	42.96	4.9	10.6	3.1	2.29	272.86
	155752		19.98	42.96	2.5	8.2	3.1	1.49	264.89
	155753		19.98	42.96	4.2	8.5	4.4	2.08	270.81
	155754		19.98	42.96	4.1	9.8	2.9	2.01	270.06
9	3b	Peralkaline granite	23.81	40.98	16.0	40.0	3.4	7.19	321.93
	1		23.80	40.95	3.0	2.0	3.2	1.21	262.14
	12a		23.83	40.94	4.0	4.0	3.2	1.61	266.09
	16c		23.82	40.93	14.0	39.0	2.6	6.54	315.39
	5a	Monzogranite	23.80	41.15	3.0	14.0	4.2	2.14	271.37
	20		23.82	41.00	3.0	21.0	3.8	2.58	275.78
	22		23.83	40.95	3.0	19.0	3.8	2.44	274.39
	39a		23.84	40.92	7.0	20.0	4.2	3.58	285.79
	4	Red granite	23.90	41.10	11.0	35.0	4.2	5.64	306.40
	31		23.90	40.90	12.0	24.0	3.6	5.08	300.78
	38		23.76	40.92	49.0	117.0	4.3	21.09	460.88
	3a	Dumah granodiorite	23.79	41.93	1.0	3.0	1.3	0.59	255.90
11	165006	Hawaiite	20.73	39.69	1.2	3.3	1.1	0.64	256.43
	175764	Gabbro	20.47	40.20	0.7	3.1	0.7	0.35	253.49
	175776	Diabase	19.38	41.25	0.2	1.7	0.4	0.22	252.22
	175785	Diorite	20.91	39.89	2.1	7.8	1.8	1.32	263.18
	165574	Qz.diorite	20.80	39.86	1.6	3.6	1.6	0.84	258.38
	165546	Rhyolite	20.74	39.89	13.8	40.6	3.3	6.22	312.20
	165571	Basalt	20.80	39.86	0.3	1.4	1.0	0.29	252.92
	165627	Cooendite	20.81	39.90	5.2	17.0	3.4	2.83	278.34
	165661	Diabase	20.95	39.68	0.9	2.7	1.0	0.51	255.12
	165881	Basalt	20.87	39.94	0.8	3.3	1.3	0.56	255.58
	165882	Hawaiite	20.90	39.90	0.6	1.2	1.0	0.32	253.24
	175780	Diabase	20.81	39.88	0.4	1.7	0.7	0.28	252.83
	165714	Perlite	20.74	39.63	3.6	7.9	3.5	1.80	268.01
	165767	Alk. basalt	20.96	39.62	0.4	1.2	0.1	0.20	252.00

2: Stuckless et al., 1986; 3: Al-Saleh & Al-Berzan, 2007; 6: Harris & Marriner, 1980; 7: Elliott, 1983; 8: Stuckless et al.; 9: Harris et al., 1986; 11: Palister, 1986.

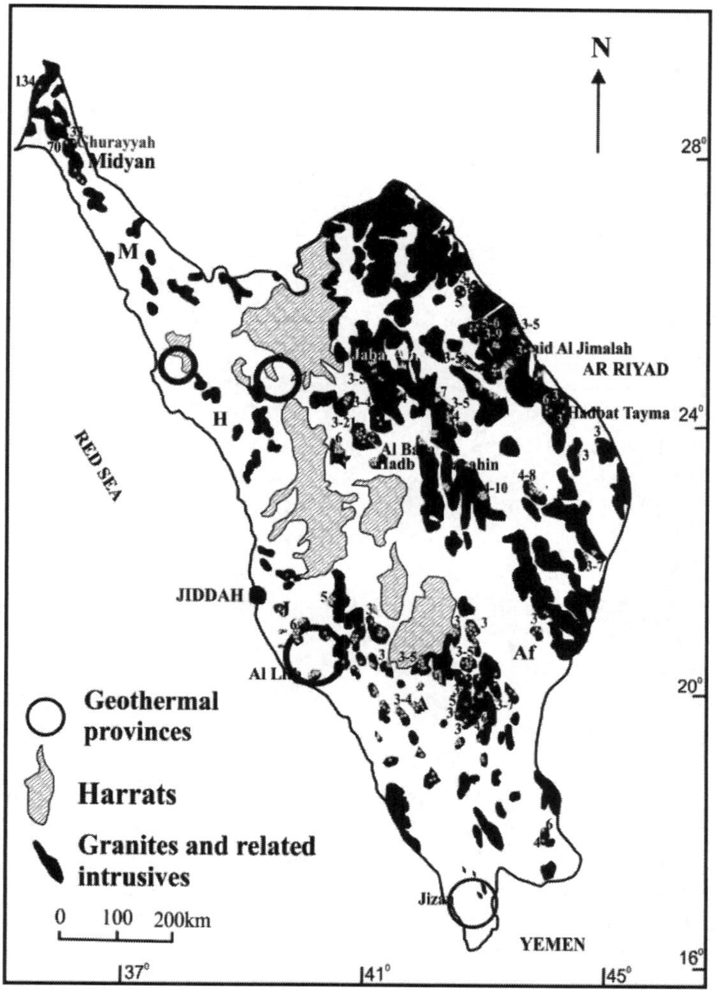

Figure 9.12 Map showing the distribution of granites and related rocks and the heat generation values of certain granites in the Arabian Shield (adapted from Chandrasekharam et al., 2015d).

The heat flow values shown in Table 9.4 are similar to the heat flow values of 175 mW/m^2 recorded in the boreholes drilled into the Precambrian basement along the western margin of the Red Sea in Egypt (Morgan and Swanberg, 1978) and over the geothermal sites along the eastern margin of the Gulf of Suez (Zaher et al., 2012). The geothermal gradient measured in these boreholes and those drilled into the Paleocene sediments varies from 40 to >80°C/km. A numerical simulation of the temperature distribution with depth at the Hammam Faraun geothermal site gave values of 170–180°C at 2 km depth (Morgan and Swanberg, 1978). In fact, some of the geothermal systems located in granites (i.e. at Al Lith, Lashin et al., 2014 and at Jizan, Chandrasekharam et al., 2015a,b) are driven by such heat generating radiogenic granites. The tritium content in the thermal waters from the above sites indicates long circulation times (>600 years, for Al Lith, Lashin et al., 2014, and 12–32 K years for Jizan, Chandrasekharam et al., 2015a,b) within the granites indicating prolonged water rock interaction giving rise to high chloride content (Al Lith: 1594 ppm, Lashin et al., 2014, Jizan 1934 ppm, Hussein and Loni, 2011) in the thermal waters.

9.3.3 Regional stress over the Arabian Shield

9.3.3.1 *Deformation stresses*

The Arabian Shield has undergone several styles of stresses since 630 Ma (Hamimi et al., 1986). The Proterozoic basement rocks are the largest tract of juvenile Neoproterozoic continental crust on earth (Figure 4.28) (Patchett and Chase, 2002). This crust has been subjected to three distinct deformation states. The D_1, D_2 and D_3 are the three initial deformation stresses that were imprinted (Hamimi et al., 2013, 2014) over the oldest lithological units (i.e. syn/post tectonic granites and related suits and members of the Ablah Group of rocks (i.e. metavolcanics and metal clastics) (Hamimi et al., 2013, 2014). These three deformation states reveal the initial stress conditions undergone by the ANS. During the D_3 deformation stage, the stress regime changed from E-W to N-S compression. Thus the Arabian Shield had been subjected to E-W and N-S compression between 786 to 630 Ma (Hamimi et al., 2013, 2014). The D_1 deformation stage was imprinted on the 689 Ma old E-W elongated ellipsoidal Mizil granitic gneiss dome in the eastern Arabian Shield (Al Saleh and Kassem, 2012). It was considered by several researchers that the Mizil granite was a metamorphic core complex and not an intrusive (structural dome) which resulted due to existing stresses pattern. However, extensive field related structural analyses, petro fabric stud and stress finite strain analysis and Ar-Ar ages reveal that the Mizil dome is indeed a diapir intruded into the volcano sedimentary pile in an island arc setting. The unusual orientation of the dome is due to a regional field stress regime which existed at 690 Ma when convergence between East and West Gondwanaland (Figure 4.28) had a NW-SE direction. This direction subsequently changed to E-W convergence due to the rotation of the Arabian plate hinged on southern Africa.

9.3.3.2 *Regional stresses*

Initial stresses that were locked in the basement rocks and associated structures can be evaluated from the Najd fault system. Between 600 and 540 Ma extensive left lateral faulting along the NW-SE trending Najd fault system cut across the Arabian Shield (Figure 9.13).

The NW-SE striking Najd fault system (Figure 9.12) is a prominent structure in the shield region. It is a strike-slip fault system with a width of about 300 km and covering a length of 1,100 km in a NW direction. Along the Red Sea coast, the Najd fault system is truncated by the Red Sea faults that developed during and after the initiation of the rift. Since the Najd fault system developed within the Arabian-Nubian Shield, the fault can be traced into the eastern desert region in Egypt. The Phanerozoic strata in the south-eastern part of the shield has also been affected due to block movement in the basement along the fault. Igneous activities during 580–530 Ma along the Najd fault system indicate that the entire system was active during this period. Aeromagnetic maps

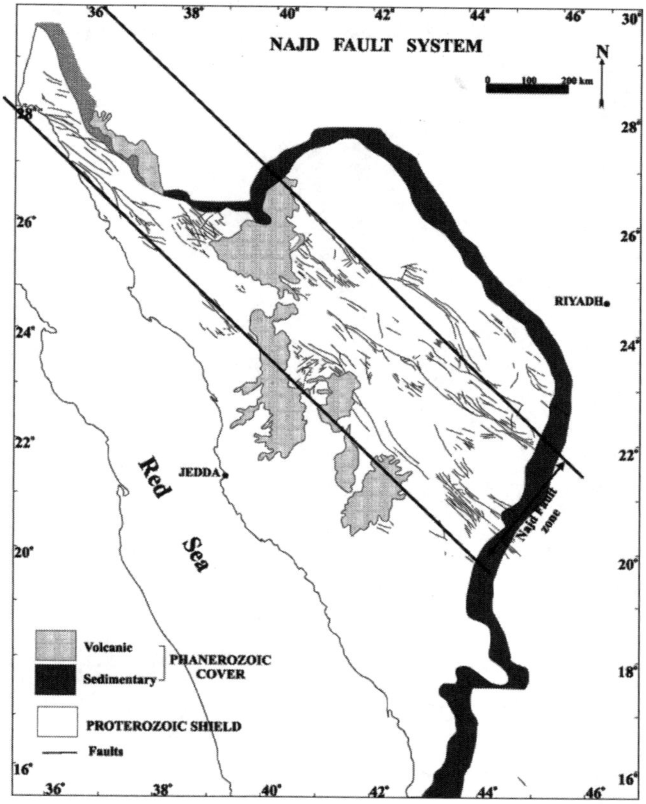

Figure 9.13 Najd Fault system, Arabian Shield.

(Figure 9.14) of the region indicate that the Najd fault system penetrates deep into the Phanerozoic cover and is a much wider subsurface. Analysis of the fault system reveals that the deviatoric stresses have been accommodated by simple shear on established faults and brittle failure on new extensional fractures. The Najd system is dominated by satellite faults striking north-west. The Najd fault system is the result of shear that allowed the Nubian and southern Arabian plates to move several hundred kilometres sinisterly with respect to northern Arabia (Stern, 1985, Moore, 1979).

In the western part of the shield, evaporite rift basins developed bounded by N-W trending faults which are parallel to the Najd fault system. Therefore, all the E-W trending stress parameters related to collision tectonics terminated between 640–600 Ma and the shield subsequently experienced NW-SE directed crustal extension related to continental, breakup (Husseini, 1988).

9.3.4 Gulf of Aden and South Red Sea stress regime

The stress regime since 30 Ma changed due to the rotation of the Arabian Plate with respect to Africa as a consequence of the opening of the Red Sea and Gulf of Aden.

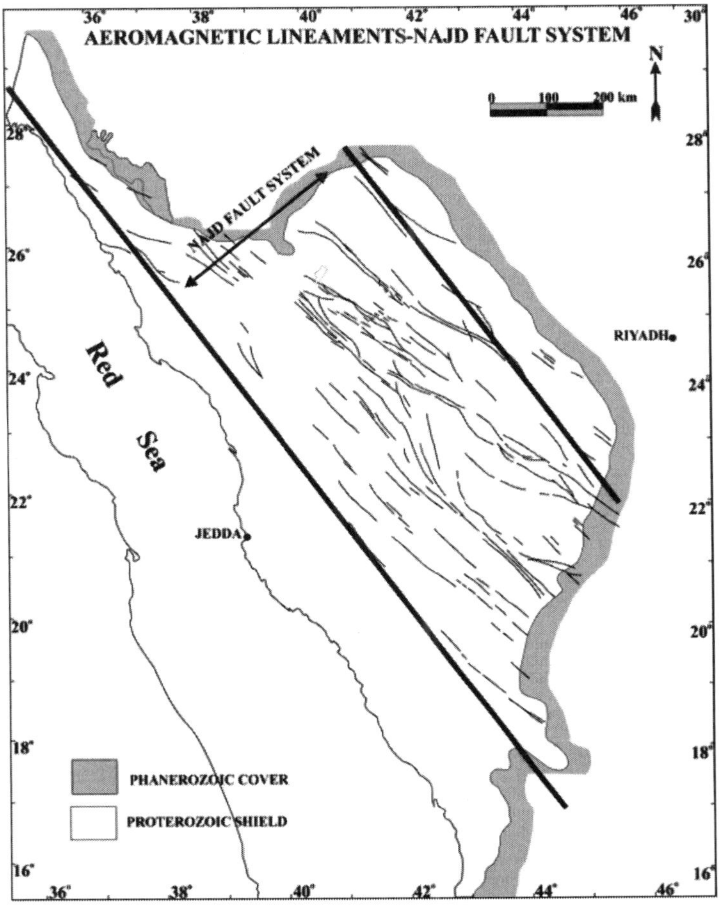

Figure 9.14 Aeromagnetic signature of Najd fault system subsurface.

The breakup of the continents was sharp and the total Red Sea floor and the Gulf of Aden were made of entirely ocean floor (McKenzie et al., 1970, Bohannon, 1986, Le Pichon and Gaulier, 1988, Bosworth et al., 2005). The initial movement at around 30 Ma started at the rate of 1 mm/yr and during the last 5 Ma this rate increased to 9 mm/yr (Bayer et al., 1988) along the E-W and SW-NE directions. Figure 9.15 gives the present day extensional, rotational, transverse and subduction related stresses operating over the Nubian-Arabian-Indian and Eurasian plates. The present stress regime commenced since the initiating of the Red Sea rift and opening of the Red Sea, which commenced initially near the Gulf of Aden and subsequently propagated towards the north. Bohannan (1989) and Bosworth et al. (2005) presented contradicting views on the initiation of the Red Sea rift, opening of the Red Sea and evolution of the Nubian-Arabian plate since the Late Eocene-Early Oligocene.

According to Bohannan (1989), the sequence of formation of the Red Sea was a) initial alkaline magmatism at about 30–32 Ma along a narrow linear zone in the

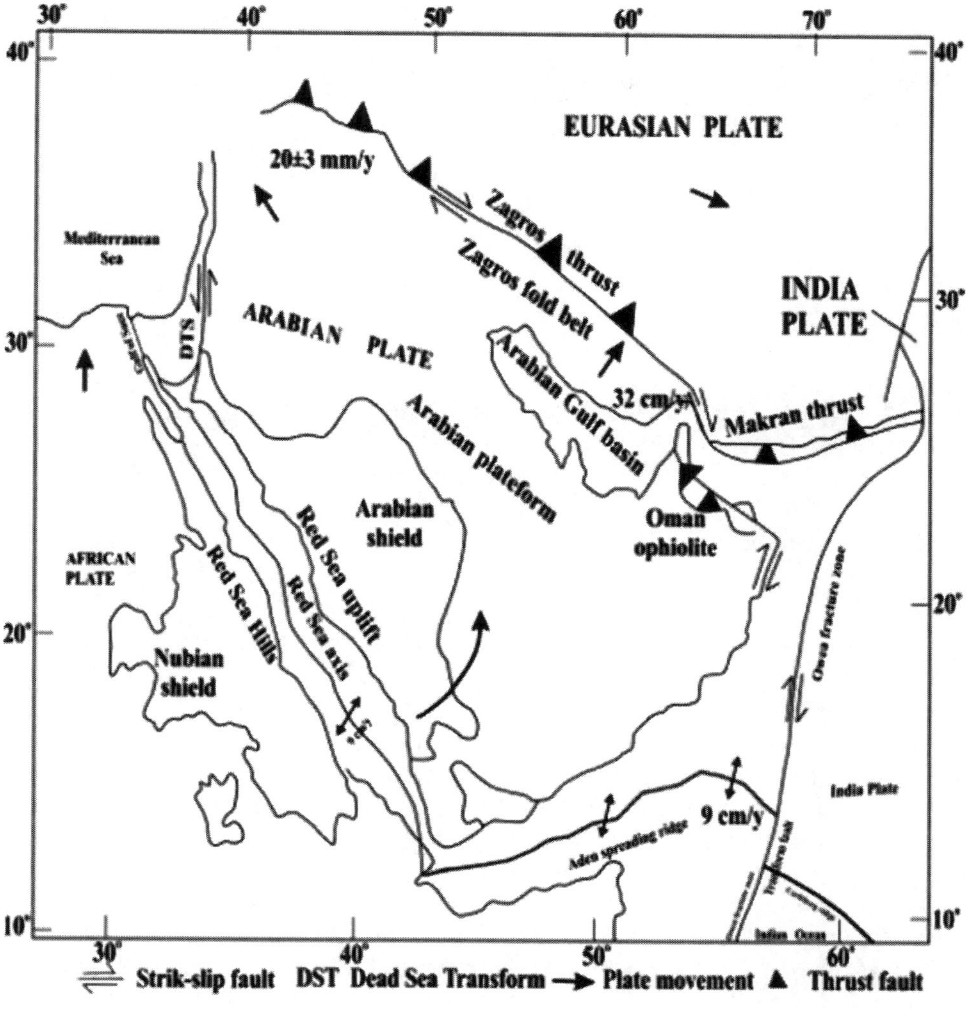

Figure 9.15 Present day stress operating over the Nubian-Arabian plate (adapted from Bosworth et al., 2015).

shield, b) rotational and detachment faulting at 25 Ma, c) gabbro and diorite intrusive and acid volcanic phase between 20 and 25 Ma with simultaneous non-marine sedimentation in the initial rift, d) deposition of marine sediments and subsidence of shelf in the Miocene and e) uplift of adjoining continental area subsidence of the shelf between 13 and 5 Ma. The above sequence of processes was caused by a lithospheric extension that is linked to the collision of India and Eurasia. Due to this collision, the Owens fracture zone north of the Carlsberg Ridge became rigid, forcing the Arabian plate to rotate along with India and move away from the African plate. As a result, the Arabian Shield was under an extension regime while the eastern margin was under compression due to the collision with the Iran plate (Figure 9.16). This initial extension

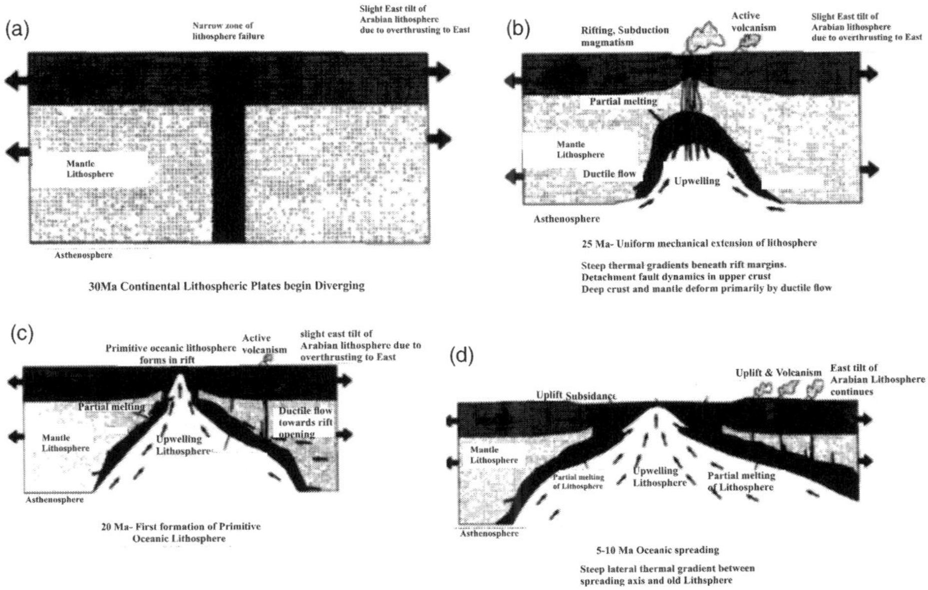

Figure 9.16 Evolution of the Red Sea rift due to slab pull (adapted from Bohannon et al., 1989).

regime must have caused intrusion of dike swarms parallel to the coast before the evolution of the harrats (Chandrasekharam et al., 2014a,b, 2015a,b). Makris and Rihm (1991) proposed a pull-apart model for the evolution of the Red Sea based on seismic, gravity and magnetic anomaly investigations. According to this model, the breakup of the African and Arabian shields was initiated by strike-slip faulting, a similar process advocated by Bohannan et al. (1989) described above. The strike-slip model advocates early oceanisation of the Red Sea due to the onset of pull-apart basins and extensive intrusive (dike) activity along the eastern coast of the Red Sea. The strike-slip process was facilitated by the inherent weakness (e.g. Najd shear system) developed by the shield during 600 Ma Pan-African orogeny (Makris and Rihm, 1991). The initial faults and lineaments that developed due to the pull-apart structural regime were subsequently reactivated during the Oligocene and resulted in extensive magmatic activity which continues even today. An in situ field stress investigation at the Mudhiq dam site in the Taif region, western Saudi Arabia (Giraud et al., 1986), in conjunction with studies of earthquake focal mechanisms, fracture patterns, and topography demonstrated that the major stress field of the shield was oriented along the NE-SW to E-W direction with the Arabian plate as demonstrated earlier, rotating at an angle of 6°C in an anticlockwise direction with respect to Africa at a rate of 9 cm/y (Girdler, 1966, Girdler and Evans, 1977, Garson and Krs, 1976).

The evolution scheme for Red Sea proposed by Bosworth et al. (2005) was entirely different from that proposed by Bohannan (1989).

Bosworth et al. (2005) introduced plume theory to explain the initial volcanism and associated tectonism that were responsible for the opening of the Red Sea and the

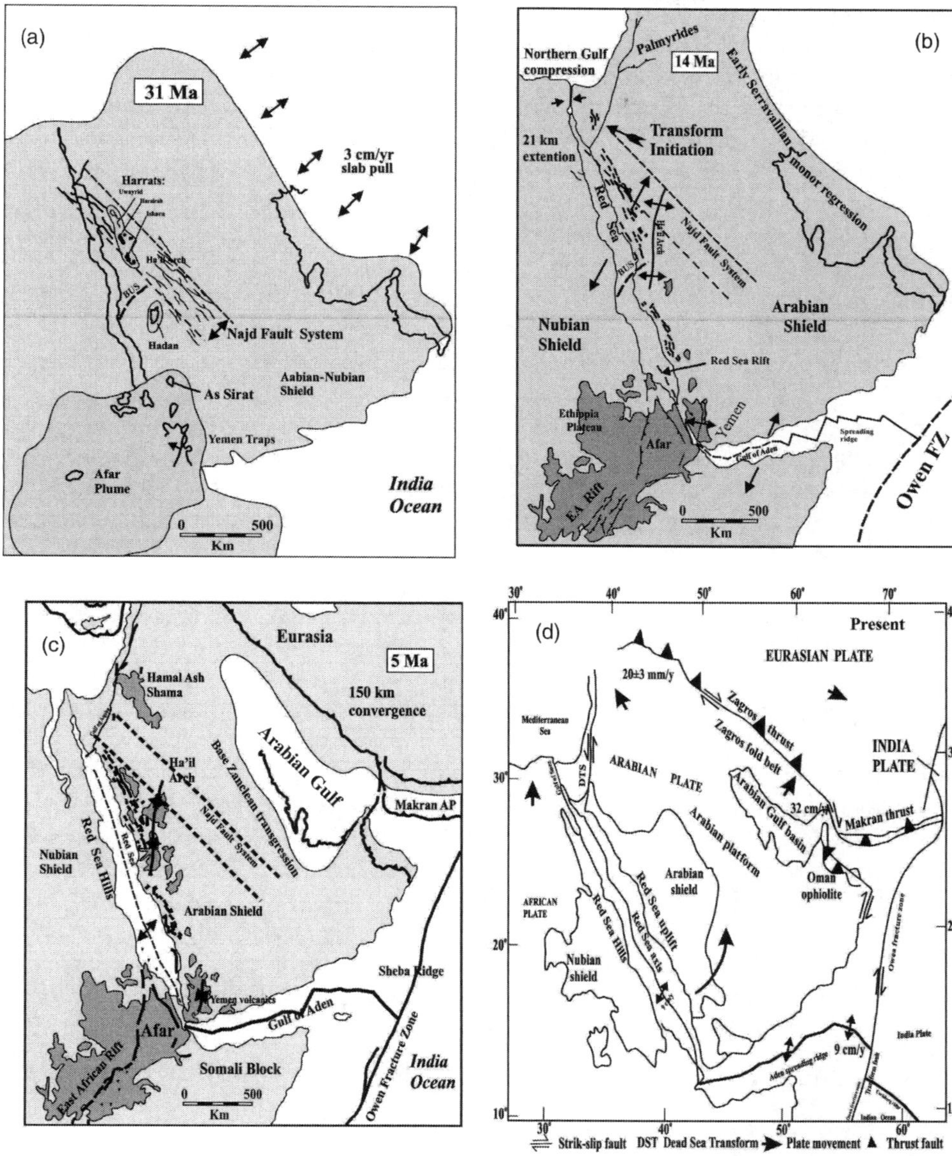

Figure 9.17 Schematic diagrams (a–c) showing the opening of the Red Sea and the formation of the rift axis and the present day (d) stress conditions over the Arabian Shield and Nubian Shield (adapted from Bosworth et al., 2005).

formation of the spreading Red Sea ridge. The initiation of the Red Sea rift began with the opening of the Gulf of Aden rift system during the Early Oligocene (~31 Ma). The entire process of rifting was caused by a plume below Afar (Figure 4.31). The sequence of events that caused the formation of the Red Sea and initiation of the Red Sea rift is shown in Figures 9.17a–d.

The initial plume formed below the Afar region propelled the Red Sea rift that started from the southern part (present Aden-Afar-Red Sea axis junction) and propagated northwards. This process started around 31 Ma with the activation of the Gulf of Aden rift (Figure 9.16a) and the eruption of large basalt flows over Yemen and Aiba basalt flows over Eritrea. Over the Arabian Shield the plume activity resulted in the formation of three prominent plume related harrats (The older harrats). They are 1) Harrat As-Sirat, 2) Harrat Hadan and 3) Harrat Uwaynd (Figure 4.32). The flows were erupted from central volcanoes and were dated at 28–26 Ma (Bosworth et al., 2005). The eruptions were strongly influenced by N-S trending arches that were formed during the Cretaceous period (Figure 9.16b). Immediately, due to the tensional stresses exerted by the plume and slab pull from the eastern margin of the Saudi Arabian platform, the western Arabian Shield was intruded by dike swarms, extending from the Yemen (present day) coast to the Midyan region (Figures 4.32 and 9.16a) which were dated at 20–25 Ma. The volcanic activity that commenced from ~30 Ma is still continuing (e.g. Harrat Lunayyir) indicated by a magma uprise associated with earthquake swarms (Hamlyn et al., 2014, Moufti et al., 2013, Duncan and Al Amri, 2013, Bayer et al., 1989, Camp and Roobol, 1992, Pallister et al., 2010, Al-Shanti and Mitchell, 1976, Koulakov et al., 2015). The spreading rate started from 1 mm/yr to the current 9 mm/yr (Figure 9.16d). With the activation of the Aquaba fault and the extension of the Suez fault and the Owen fracture zone which has been connected to the Gulf of Aden spreading zone, the Saudi Arabian plate started rotating anticlockwise with the tensional forces acting perpendicular to the rift axis with a direction towards NNE-SSW. During the anticlock-wise rotation period, structures like the Najd fault system was reactivated during the entire period of this tectonic evolution. The stretching factor of the continental crust was about $\beta = 1.15$ over the Gulf of Aden and central Yemen, 1.6 to 1.8 onshore Red Sea near Yemen, 1.6 south of Gulf of Suez, 2.4 over the Red Sea margin near Yemen (Bosworth et al., 2005). The present day stress field of Arabia and Africa is decoupled with the increase in the intensity of East African-Kenya rifting. The maximum horizontal stresses in Arabia are oriented along N-S and E-W in Egypt (Bosworth et al., 2015). Whatever the process of evolution of the Red Sea (Bosworth et al., 2005, Bohannon et al., 1989) and the changes of stress directions that the shield has undergone since 700 Ma (Makris and Rihm, 1991), the high heat generating granites of the western shield are under a compressional regime along the NE-SW to E-W direction.

9.3.5 Stress regime around the Gulf of Suez

The Gulf of Suez tectonic style evolved subsequent to the southern part of the Red Sea rift initiation in the Late Oligocene. Extensional fault activity started around 27 Ma in the northern part of the Red Sea rift towards the southern part of the Gulf of Suez. Fault geometry, fault kinematics and sedimentation style suggest that a NE-SW extension played a major role in the rift initiation. The activation of the basement structures (e.g. like the Najd fault system) resulted in pull-apart basins. The extension rates of the Gulf of Suez retarded from an initial stretching factor of $\beta = 2$ (Gaulier et al., 1988) due to the linking of the Levant Gulf Aqaba transform fault with the Zagros plate boundary, which caused the Arabian plate to rotate anticlockwise.

Unlike the southern Red Sea basement, the Gulf of Suez basement consists of crystalline Precambrian granitic gneiss with NW trending fabric, which is an imprint of the

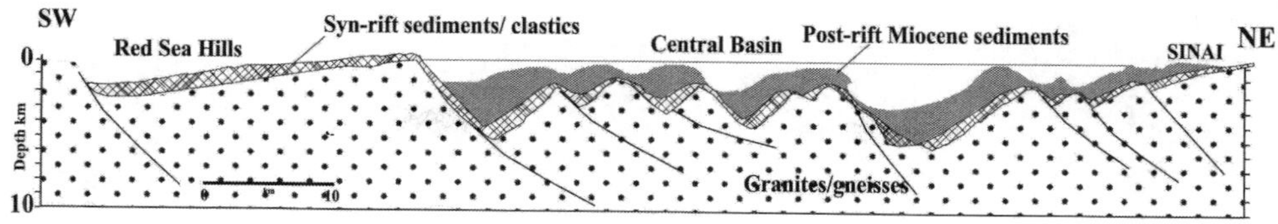

Figure 9.18 Subsurface lithology between the Red Hills (see Figure 4.12 for the cross section line). Unlike the southern Red Sea, the Gulf of Suez has Precambrian crystallines (Nubian shield granites and gneisses). The Sinai and Red Sea coasts are the loci of several thermal springs (see Figure 4.2) and also the zones of high heat flow (see Figure 9.9). The rift was initiated during Oligocene (adapted from Bosworth and Mc Clay, 2001).

Najd fault fabric (Figure 9.16c). The basement rocks are similar to those exposed along the eastern desert region, described in the above section. The subsurface lithological sequence across the Gulf of Suez (SW-NE section in Figure 4.2) is shown in Figure 9.18.

The Gulf of Suez rift is one of the best examples of a continental rift formation and extensional fault development which represents large scale fragmentation along the axis forming sub-basins (Gawthorpe et al., 1997). The paleo-stress fields that played a significant role in the evolution of the Gulf of Suez rift have been analysed by several researchers: Angelier, 1985, Bosworth and Tavian, 1996, Bosworth, and Mc Clay, 2001, Gawthorpe et al., 1997 Jarrige et al., 1990, Lyberis, 1988). The paleo-stress pattern proposed by this research for the Gulf of Suez is shown in Figure 9.19. It is apparent from Figure 9.18 that extensional stress regime was prevalent between 25 to 5 Ma and the pattern changed subsequently due to the style of drift and rotation of the Arabian Shield.

9.3.6 Heat flow and subsurface temperature of the Arabian Shield

The radioactive heat production (RHP) of the granites, measured in $\mu W/m^3$, has been calculated using the equations of Raybach (1976) and Cermak et al. (1982) taking into account the heat generation constant (amount of heat released per gram of U, Th and K per unit time) and the concentrations of U, Th and K (C_u, C_{Th} and C_K) (see Table 9.4):

$$RHP = \rho(9.52C_U + 2.56C_{Th} + 3.48C_K) \times 10^{-5}$$

where ρ is the density of rock in kg/m^3; C_U and C_{Th} are the concentration of U and Th in mg/kg respectively and C_K is the concentration of K in weight percentage in the granites (Table 9.4). The heat generation values by the granites from a selected area over the western shield region are shown in Figure 9.11 (Only values $>3\,\mu W/m^3$ are shown).

The surface heat flow values were calculated using the following equation of Lachenbrough (1968):

$$Q = Q_0 + D \times A$$

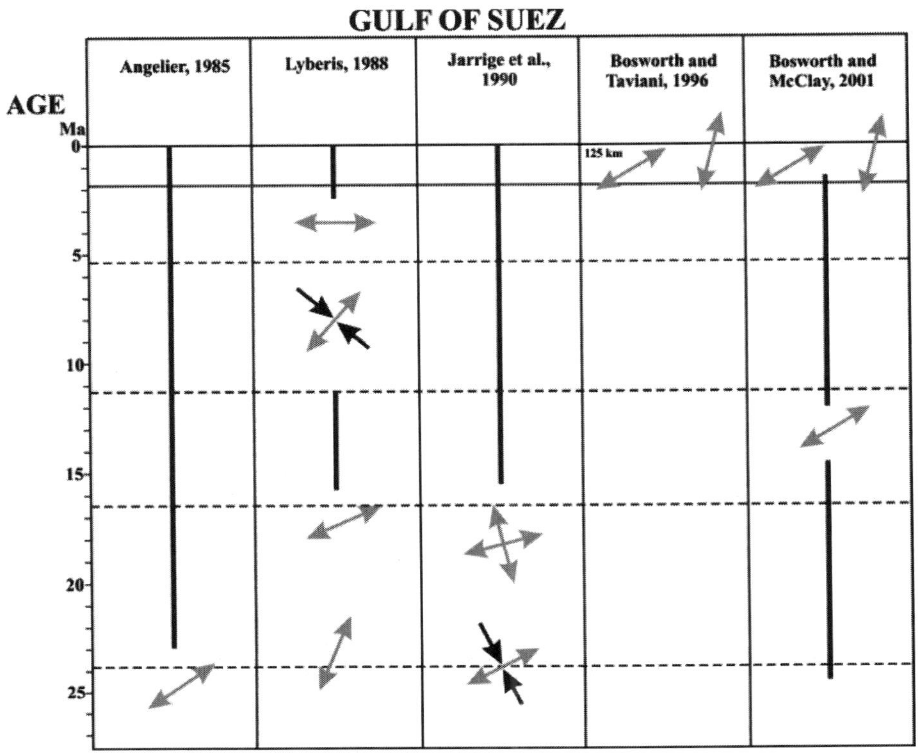

Figure 9.19 The interpretation of paleo-stresses deduced by structural restoration of tectonic events and fault kinematic data that resulted in the opening of the Gulf of Suez between 25 Ma and the present (adapted from Bosworth and McClay, 2001).

where Q is the heat flow at the surface, Q_0 is an initial value of heat flow unrelated to the specific decay of radioactive element at certain time, D is the thickness of rock over which the distribution of radioactive element is more or less homogeneous, and A is the radioactive heat production. Since the Moho depth along the western Arabian coast varies from 18 to 25 km (Park et al., 2008), the background heat flow (Q_0) value along the shelf is considered as 250 mW/m^2 and along the coastal region (between the shore line and the escarpment) the value is taken as 100 mW/m^2 (Girdler, 1977). The heat flow values thus calculated over the shield are listed in Table 9.4. Based on the heat flow values (Table 9.4), the subsurface temperatures have been calculated using the following relation (Vernekar, 1975):

$$Q = k(dT/dZ)$$

where k is the thermal conductivity of the rock and dT/dZ is the thermal gradient. The subsurface temperature has been calculated by taking the average surface temperature as being 30°C (Vernekar, 1975) and thermal conductivity of the rock (i.e. granite) as 3.98 Wm^{-1}C^{-1}. The estimated subsurface temperature of the Arabian Shield at 2.5 km depth is shown in Figure 9.19. The heat conduction due to the volcanic activities

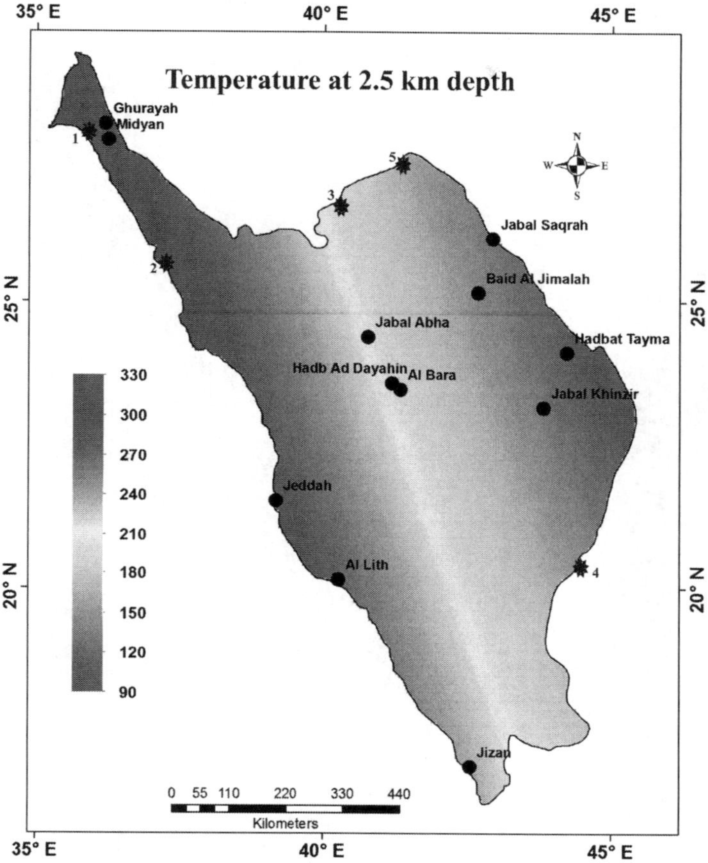

Figure 9.20 Subsurface temperatures of the Arabian Shield at 2.5 km depth. Temperature is calculated based on the heat generation of the granites and the heat flow values shown in Table 9.4. Stars represent borehole locations (see text) (adapted from Chandrasekharam et al., 2015d).

(harrats) is not incorporated in the preparation of the subsurface temperature shown in Figure 9.20. If this is incorporated, then there will be large high temperature anomalies around the harrats as shown by Khyber and Harrat Lunayyir (see Figure 4.33).

9.3.7 Gamma ray logs in bore wells

The natural gamma ray log(API) together with the resistivity, thorium, uranium and potassium content and lithology have been recorded from 5 bore wells in the Arabian Shield region. The location of these bore wells are shown in Figure 9.19. The heat generated by the rocks within the boreholes was calculated using the relationship suggested by Becker and Rybach (1996) and is shown below:

$$RHP = a(\mathrm{GR[API]}) \pm b$$

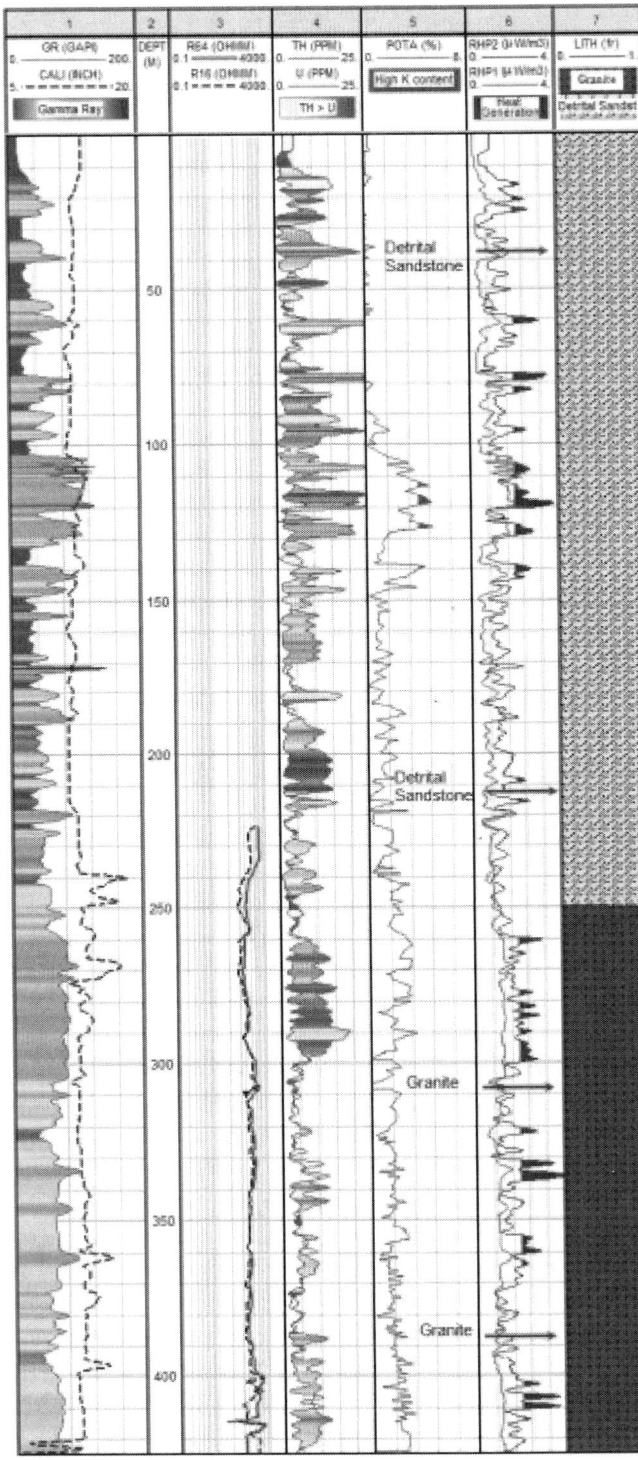

Figure 9.21a Gamma ray spectrometry logs for bore hole 1 (see Figure 9.19 for borehole location) (adopted from Chandrasekharam et al., 2015d).

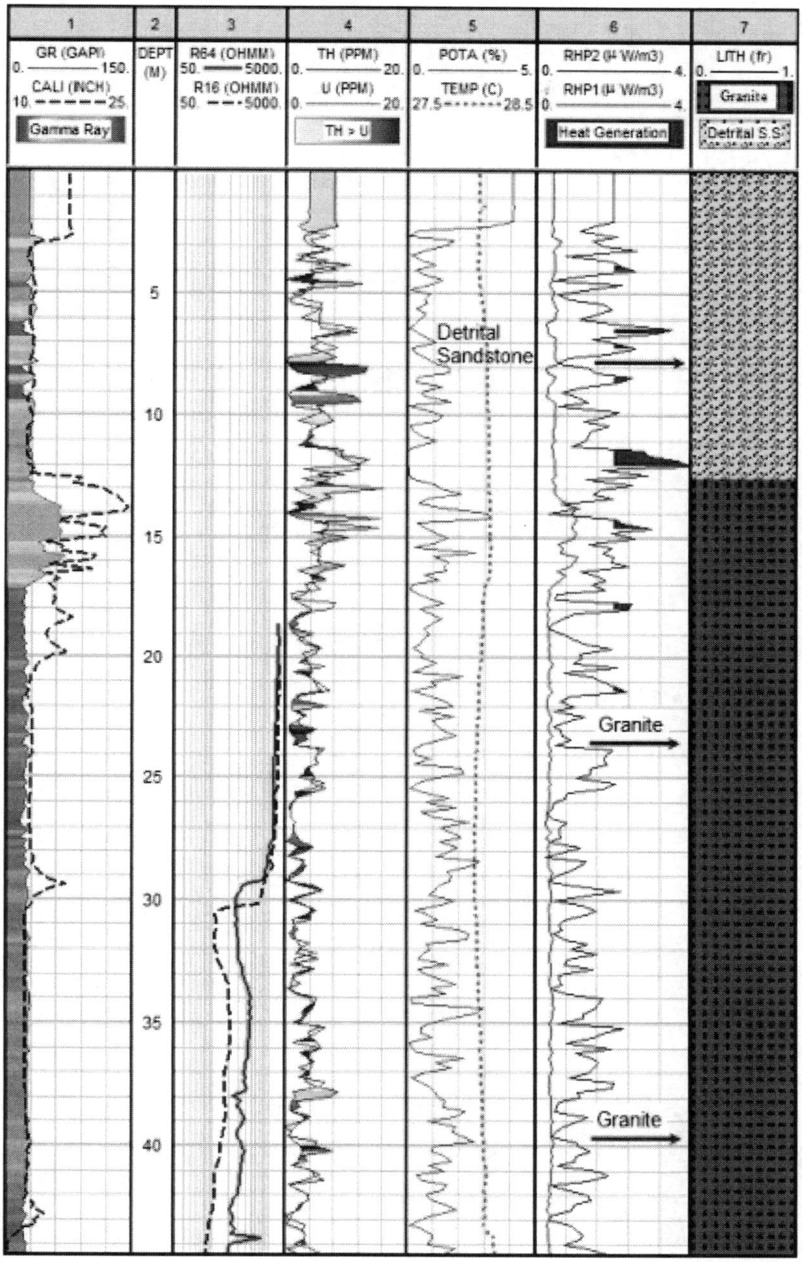

Figure 9.21b Gamma ray spectrometry logs for borehole 2 (adopted from Chandrasekharam et al., 2015d).

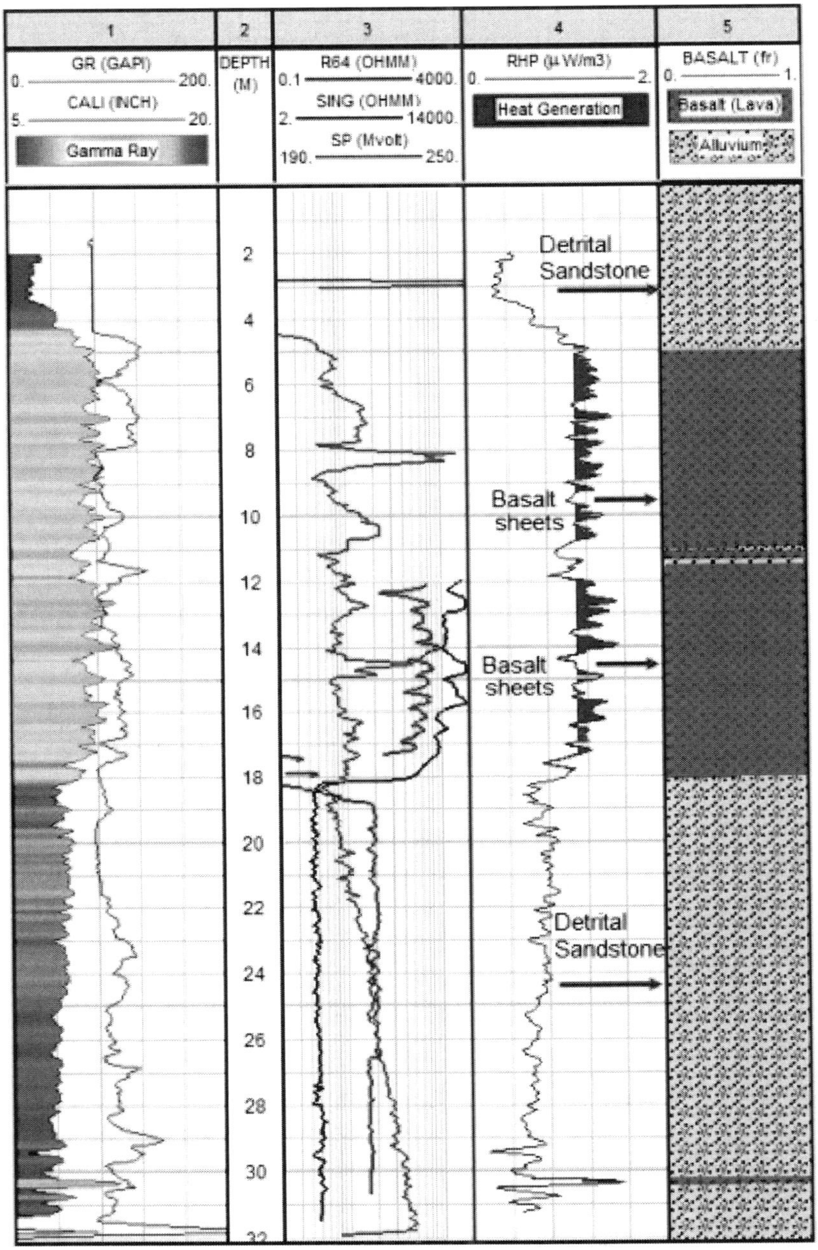

Figure 9.21c Gamma ray spectrometry logs for borehole 3 (for borehole location see Figure 9.19) (adopted from Chandrasekharam et al., 2015d).

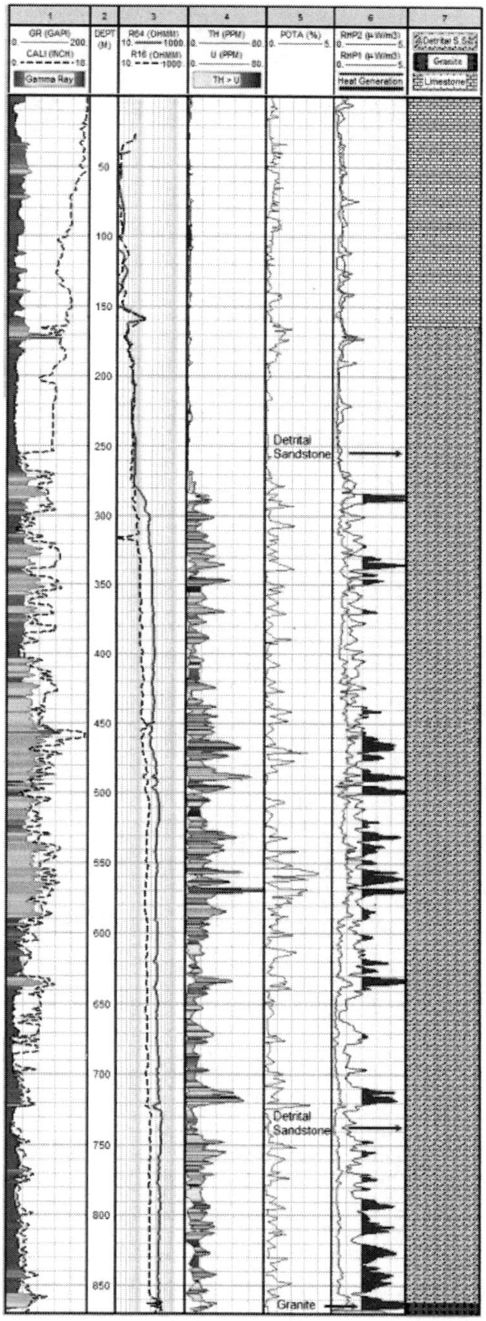

Figure 9.21d Gamma ray spectrometry logs for borehole 4 (adopted from Chandrasekharam et al., 2015d).

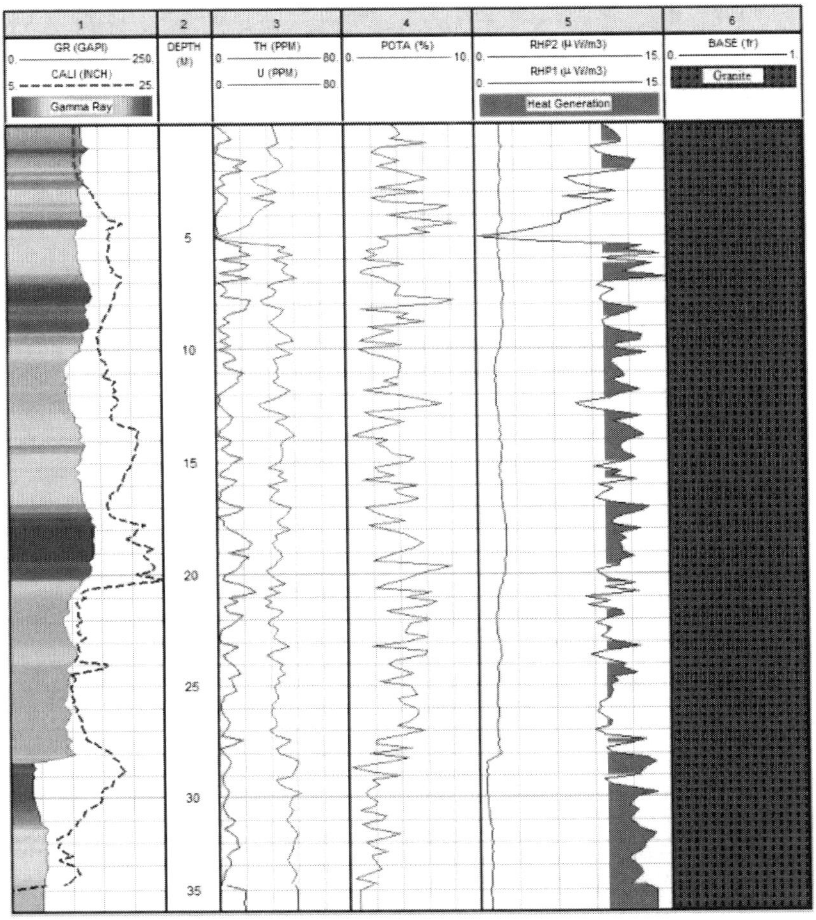

Figure 9.21e Gamma ray spectrometry logs for borehole 5 (for borehole location see Figure 9.20) (adopted from Chandrasekharam et al., 2015d).

where RHP is the rate of radiogenic heat production ($\mu W/m^3$), and *a* and *b* are empirical constants. The natural gamma ray logs of the five bore wells, shown in Figure 9.19 are shown in 9.21a to c.

The lithologs show a thin sedimentary cover (detrital sandstone) over the hard crystalline basement granites (Figures 9.21a, b, d). In borehole 4 (Figure 9.21d), located towards the SE periphery of the shield, thin carbonate rocks are present at near surface depth. It is interesting to note the high level gamma ray signature of the detrital sandstone, with a heat generation value of 2–4 $\mu W/m^3$, and with a high content of thorium (Figures 9.21a, b, d). The high radio activity in the detrital sandstone appears to be due to the derivation of the detrital material from the granitic rocks. The high caliper log response (20–25 inch) at the contact between the detrital sandstone and the underlying granite rocks indicates the highly fractured nature of the upper surface of the granitic rocks (Figures 9.21a, b).

The high radioactive heat production (RHP > 5 $\mu W/m^3$) associated with the high gamma ray response is recorded in the middle (see Figure 9.20a) for well 1 and the lowermost part of the granitic rocks (Figure 9.20d, well 4). Low heat generation (1–2 $\mu W/m^3$) is represented by the thin basaltic sheets which are frequently interbedded with the detrital shallow sandstone (Figure 9.21c).

Very high heat generation is recorded where the granites are exposed at the surface (Figure 9.20e, well 5). The well is a shallow one, located at the extreme northern borders of the Arabian Shield (very close to Hail province), and is drilled into 35 m thick granite. The heat generated by these granites is around 10–15 $\mu W/m^3$. The abnormal high response of the caliper log (>20 inch) in the middle of the granite section indicates the highly fractured state of the granite. The heat flow values recorded in the bore holes and those calculated using the radioactive elements (Table 9.4) in the outcropping granites are similar. Since the granitic rocks were evolved when the Nubian and the Arabian shields were together, these rocks on either side of the Red Sea (Nubian shield in Egypt and Arabian shield in Saudi Arabia) have similar heat generation capacity and the heat flow values over the surface are similar (see section 9.3.1). Although data on the uranium and thorium concentration in granitic rocks occurring in Eritrea and Yemen are not available, the heat production and heat flow values of the rocks occurring in Yemen and Eritrea will be similar to those recorded in the Arabian and Nubian shields.

Chapter 10

Economics

Today, over two billion people in developing countries live without any electricity. They lead lives of misery, walking miles every day for water and firewood, just to survive. What if there was an existing, viable technology, that when developed to its highest potential could increase everyone's standard of living, cut fossil fuel demand and the resultant pollution.

Peter Meisen, President, Global Energy Network Institute (GENI) (1997)

The most important economic aspect of geothermal energy, besides power, is the cost benefit. Countries need not depend on imported fuels thereby saving enormous amounts of money. While in the case of conventional energy sources, the source of energy has to be bought, which has tremendous effects on the power generation support systems – from generation to supply costs. Besides savings related to the fuel cost, geothermal energy can earn additional revenue through carbon savings. However due to prevailing tax structures in different countries, the cost of a geothermal project varies and hence the cost of power. This is true especially with respect to non-OECD countries which are using obsolete technologies to generate power and not adopting clean development mechanisms to mitigate CO_2 emissions. Geothermal power projects have two major costs: capital cost: this includes the cost of the land or land lease cost, surface and subsurface exploration costs, site infrastructure development cost, power plant cost and 2) maintenance cost. Further, the composition of the geothermal fluids and the depth of the geothermal reservoir, which are site specific, and the temperature of the reservoir have a direct bearing on the cost of the project. High-enthalpy fluids and shallow reservoir depth reduce the cost of the project compared to high-enthalpy fluids and deeper reservoir depth. Now that geothermal technology is very mature, exploration and exploitation of low-enthalpy fluids at shallower depths are very cost effective (Chandrasekharam and Bundschuh, 2008).

10.1 LAND AND EXPLORATION COST

Amongst all the renewable energy sources, the land requirement for geothermal power plants is the minimum. The land required for a 1 MWe geothermal power plant is 1.2 acres while a wind and solar power plant require 65 to 12 acres of land to generate the same quantity of electrical power (Chandrasekharam et al., 2014a). If the land is owned by the government, then this land can be given for development through an appropriate scheme. Since most geothermal power plants are owned by governments, this is possible. In case the resources are located in a privately-owned property then an appropriate agreement may be reached between the developer and the land owner. But in several countries, it is not still clear whether all the natural resources that occur within a private property belong to the owner of that property or they belong to the government. In many countries geothermal manifestations occur in private land and inside national parks, and in such cases it is difficult to acquire the land or get the land on lease for development. But the geothermal reservoir need not lie just below the surface manifestation. Only subsurface exploration, supported by geophysical techniques, can reasonably locate the subsurface reservoir (see section 6.3). The cost incurred in surface exploration is very small compared to the cost incurred for drilling. Several published reports together with the data generated from field investigations will give substantial information on the subsurface conditions prevailing in a geothermal province (see section 6.2.1). The cost of exploration methods for a 100 MWe geothermal power project has been estimated to be US$ 7.7 per kW (Hance, 2005). But such estimates are not universal and heavily depend on the cost of the local labour and analytical cost to analyse the samples of rocks and water etc. Certain geothermal developers include the exploratory drilling cost in the exploration budget. In such cases, the cost of 1 kW power will be 40 to 50 times greater than the above mentioned kW cost. The site infrastructure development cost varies with the location of the project. If the location of the project is at a remote place, then the accessibility to the site increases the exploration cost. Thus, the exploration costs vary and depend on several factors mentioned above. Further, if the geothermal field is a green field then the cost of exploration will be higher than the cost of expansion of the existing field. The exploration cost estimates reported in the literature vary from 88 to 142 US$/kW (Sanyal, 2004), 101 to 130 US$/kW (Simons et al., 2001) and 107 US$/kW (Nielson, 1989). These costs may be considered as indicators and need not be the actual cost of exploration. Time delays in exploration (due to obtaining exploration licences, bureaucratic hurdles, especially in developing countries) are a major concern in commencing a project. This will affect the cost of the project to a large extent. As shown in the Figure 10.1, a US$ 100 capital cost project if delayed for a period of 10 years, will cost US$ 481. This increase will either reduce the profit margin or will increase the selling cost of power. The later part if not generally acceptable by the consumer then it will lower the commercial viability of the project if the project is financed by a bank.

10.2 DRILLING COST

In any geothermal project, drilling takes away 40% of the capital cost. Out of this, half of this cost is related to the time charges of the drilling rig as well as the associated equipment and manpower. If sufficient drilling equipment is not available for geothermal projects, it will increase the time delays in drilling the wells. This will apparently increase the cost of production of electricity. In addition, the drilling cost depends on the well design, depth of drilling, efficiency of the drilling equipment, breakdown period and the experience of the drilling personnel. The major cost components in drilling a well are pre-spud drilling (includes, daily operating cost, drilling

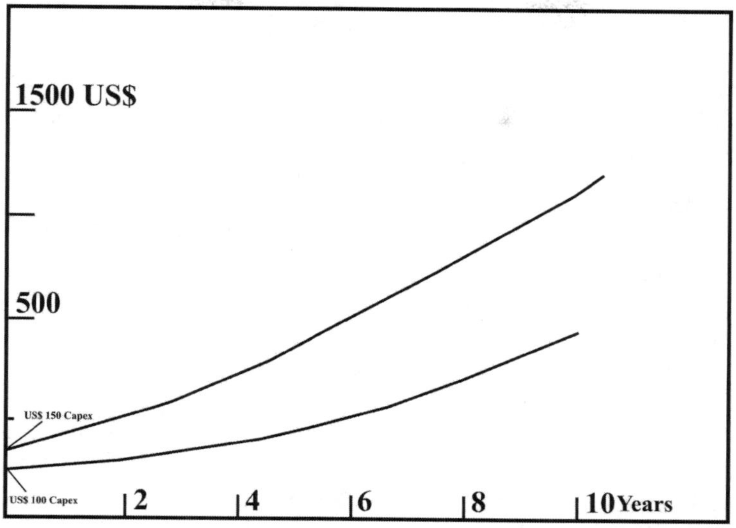

Figure 10.1 Escalation of capital cost of projects due to time delays (adapted from Hance, 2005).

Table 10.1 Cost of drilling a geothermal well (Source: Kipsang, 2013).

	Cost in US$
Pre-spud costs	400,000
Drill site preparation	400,000
Daily operating cost	3,242,700
Drilling consumables	817,900
Casing and well head	535,800
Services	99,000
Total	5,495,400
Grand Total including 10% contingency	6,045,000

consumables, well casing and well head assembly, services and non-productive costs that include problems related to drill hole) and well completion costs. An example of drilling cost with different components is shown in Table 10.1.

In Table 10.1 the daily operating cost includes rig rental with crew and standby crew, air compressor, cement equipment crane charges, water supply and disposal facility and catering to the crew and the drilling personnel. The drilling consumables include drill bits, drilling mud, diesel, additives etc. The well and well head accessories include for example the well casing blow preventer. The above example is for a well drilled in Olkaria geothermal field as reported by Kipsang (2013). Similarly, Thorhallsson and Sveinbjornsson (2012) presented statistical cost analyses of drilling based on 72 high temperature large diameter production wells drilled to 2,175 m and injection wells in Iceland between 2001 and 2009. The cost calculations were made based on market prices prevailing during that period. The average cost for drilling by large diameter well to 2,175 m is shown in Table 10.2.

Table 10.2 Average cost of drilling a 2,175 m large diameter well (adapted from Thorhallsson and Sveinbjornsson, 2012).

	US$	%
Site, water supply, cellar	400,000	8.5
Moving small and big rigs	361,000	7.8
Drilling, casing, slot production liner from 26″ to ~9″	390,7,000	83.7
Total	4,668,000	100

%: Percent of total cost.

The above authors carried out Monte Carlo simulations using probability distributions for the uncertainties in the number of working days, unit cost of material and rates for the drilling rigs. The results obtained from the simulations lie within 95% confidence level of the cost shown in Table 10.2. Sensitive analyses indicate that most of the uncertainties are related to the number of working days. These costs are applicable where the drilling and associated activities are executed without any hindrances. Nonetheless, in any drilling activity, there are always unforeseen situations which arise due to geological problems that cause additional costs, for example drill bits getting stuck in the rocks. The project should be prepared to meet such unforeseen contingencies by keeping a certain percentage of contingency costs factored in the capital cost.

10.3 POWER GENERATION COST

The ultimate success of any geothermal project depends on the power generated by the wells. If the wells are not able to generate the estimated power, then all the calculations and estimations related to the different components of the projects will lead to the failure of the project. Such situations generally will rarely arise because several checks and balances are factored into the project proposals based on a series of exploratory tests and analyses. As mentioned above, the main cost involved in geothermal power projects is drilling. The cost of the project, hence the drilling cost, decreases with an increase in the power production capacity of the well (Figure 10.2).

In the case of low-enthalpy wells, the cost of drilling can be reduced by drilling fewer production slim hole wells (Chandrasekharam and Bundschuh, 2008). Such slim holes tapping low-enthalpy resources can generate 1 to 2 MWe. Such wells are best suited for rural electrification by a dedicated local network. Geothermal wells generating 1 MWe in rural areas can support nearly 5,000 to 6,000 families (Vimmerstedt, 1998, Chandrasekharam and Bundschuh, 2008).

For power plants generating >5 MWe, although the surface cost is a smaller part of the capital cost of the project, it is linked to the amount of power generated from a well. The major cost involved in large projects is the subsurface cost. In the case of the 20 MWe Namafjall power project in Iceland, for example, the subsurface cost amounted to 37% of the total cost of the project. This is because of the large number of production wells needed to maintain the required flow rate (Stefansson, 2002). When the number of production wells increases, the number of injection wells also needs to be increased. The surface cost analysis of two high temperature wells in

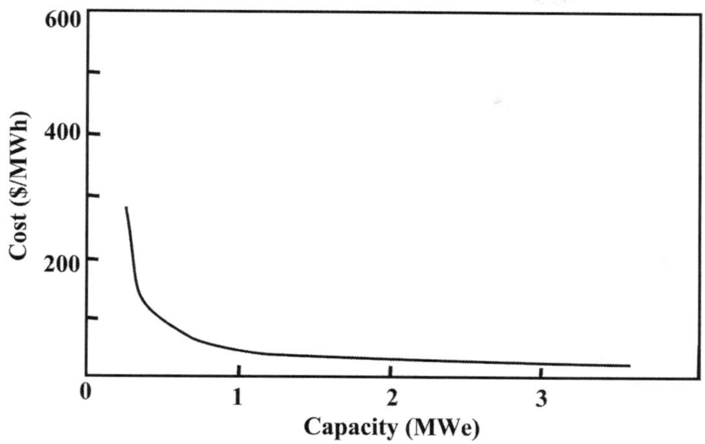

Figure 10.2 Relationship between cost of power and electricity generated by a well (adapted from Chandrasekharam and Bundschuh, 2008).

Iceland (Bjarnarflag and Krafla wells) indicates that the surface cost and size of the power plant define a linear relationship, (with a correlation coefficient, $R^2 = 0.97$) defined by the equation (Stefansson, 2002):

$$\text{Surface cost (million US\$)} = (-0.9 \pm 4.6) + (1.0 \pm 0.1) \times \text{MWe}$$

This relationship may not be applicable to all the geothermal fields but this may be considered as a guide to estimate the surface cost for geothermal power plants with capacity between 20 and 60 MWe. Large power projects (>5 MWe) also need a reasonable number of re-injection wells. The cost of the re-injection wells, in the case of larger projects, is part of the capital cost. In the case of small power plants, the cost of the injection wells is not an important component of the project cost.

Stefansson (1992), based on 31 high temperature geothermal fields located in a wide geographic locations, estimated the subsurface cost of a geothermal project. The analysis shows that the cost is more or less similar to all the projects of the world. However, in the beginning the flow rate of a production well may decrease or increase over a period of time due to the wells drawing fluids from the same aquifer. In case there is a decrease in the flow rate, cost provision should be factored into the capital cost for drilling additional wells to maintain the flow rate and the power production. However, over a period of time, the flow rate will stabilise, as shown in Figure 10.3.

The average power output from a single well is about 4 to 5 MWe for each drilled kilometre. Thus the cost estimate for the installed MWe should be made after the well stabilises and produces uniform fluid and or steam rate. An amount of US$ 37 million for 20 MWe plant or US$ 1,750 per installed kW was reported by Stefansson (2002) for a power plant in Iceland.

Stefansson (2002) further analysed the yield of the well per drilled kilometre from the above said 31 high temperature wells and presented the data in a histogram (see Figure 10.4). The actual yield is presented in Table 10.3.

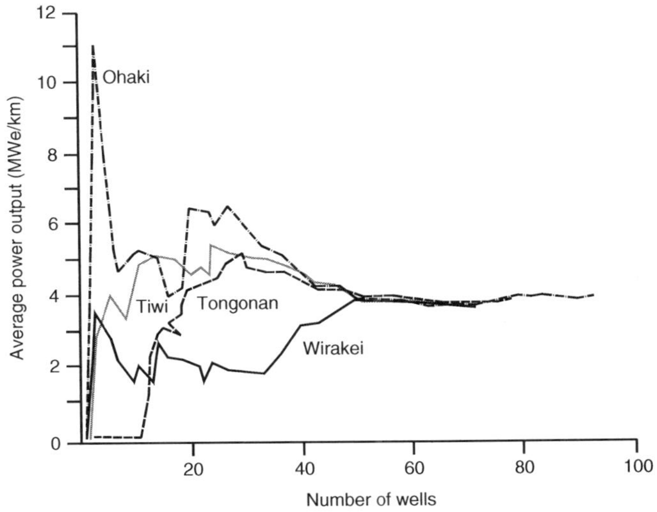

Figure 10.3 Average power output from geothermal wells from different geothermal fields (adapted from Stefansson, 2002).

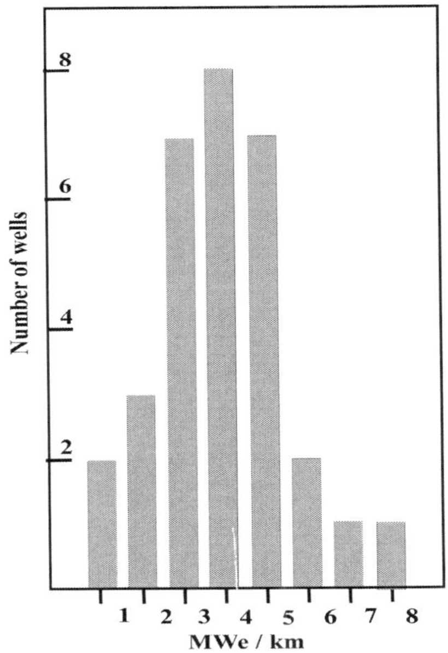

Figure 10.4 Histogram showing the yield of wells per km drilled (adapted from Stefansson, 2002).

Thus the cost estimate for a 1 MWe power project is about US$ 0.29 million from a well drilled to 1.5 km. For a 40 MWe plant, including surface and subsurface costs, the cost would be anywhere between 45 and 79 million US dollars. This works out to be on average 1,600 US$/kWe (Chandrasekharam and Bundschuh, 2008).

Table 10.3 Average production of high temperature wells (Stefansson, 2002).

Average MWe per well	4.2
Average MWe per drilled km	3.4
Average number of wells before maximum yield achieved	9.3

Table 10.4 Energy sources in rural areas across the world (adapted from Chandrasekharam and Bundschuh, 2008, WB, 1996).

	Household income		
Use	*Low*	*Medium*	*High*
Cooking	Wood, dung	Wood, dung, kerosene, biogas	Kerosene, LPG, coal
Lighting	Kerosene	Kerosene, and gasoline	Kerosene, diesel, and gasoline
Space heating	Dung (often none)	Wood, bio-residues, and dung	Wood, residues, dung, and coal

10.4 LOW-ENTHALPY GEOTHERMAL PROJECTS

A large number of geothermal sites, which are not associated with active volcanoes, produce fluids that are suitable to generate power <5 MWe (Lund and Boyd, 1999, Vimmerstedt, 1998, Chandrasekharam and Bundschuh, 2008). The reservoir temperature of such fluids is around 150°C and such geothermal resources occur in almost all the continents. These low-enthalpy systems can generate power using ORC or Kalina cycle technology and are best suited for providing power to rural communities through a local transmission network. Such systems can provide clean energy to rural communities (per capita demand of electricity in such areas is about 0.5 kW, Chandrasekharam and Bundschuh, 2008) that still use traditional fuels like kerosene, wood, cow dung, biogas, coal and agricultural produce (Table 10.4). High income families in these communities use diesel for lighting purposes.

The cost of power generated by a 2 kW diesel generator is about 1300 US$/kW, with fuel costs amounting to 0.135 US$/kWh and the cost incurred for its operation and maintenance per year is 582 US$/year (Chandrasekharam and Bundschuh, 2008, Vimmerstedt, 1998). The diesel cost is driven by fuel prices and transportation costs. Therefore, the unit cost of power keeps fluctuating with fuel and transportation costs and cannot be controlled. In the Caribbean and Latin American countries, which have large low-enthalpy geothermal resources, the average cost of geothermal power is about 25 cents/kWh (Chandrasekharam and Bundschuh, 2008, Vimmerstedt, 1998). The unit cost of power generated from low-enthalpy geothermal resources is shown in Table 10.5.

Thus, the generation cost from small geothermal power plants for rural areas is very cost effective and when such plants are linked to direct application industries, like greenhouse cultivation, dehydration and aquaculture, then the cost of power can be lowered further (Chandrasekharam and Bundschuh, 2008).

Table 10.5 Generation cost vs. resource temperature of low-enthalpy geothermal projects (adapted from Di Pippo 2008).

	Capital cost US$/kW			
	Resource Temperature (°C)			
Net power (kW)	*100*	*120*	*140*	*O & M cost US$/year*
100	2786	2429	2215	21010
200	2572	2242	2044	27115
500	2357	2055	1874	33446
1000	2143	1868	1704	48400

10.5 ECONOMIC ADVANTAGE OF GEOTHERMAL FOR RED SEA COUNTRIES

10.5.1 Egypt

The countries around the Red Sea have two pressing present and future issues: energy (electricity) and water. While counties like Saudi Arabia and Egypt have oil and gas resources, Eritrea, Ethiopia, Djibouti and Republic of Yemen depend on oil and gas imports to meet energy demand (Chapter 2). All the countries however have a common demand – fresh water for human consumption and agricultural production. These countries heavily depend on desalination of seawater to meet this demand. In the following sections geothermal energy sources' role in meeting the above demands and improving the economic status of some of the countries around the Red Sea is briefly discussed.

Egypt, with a population of 82 million, is the largest oil and gas producing country in north Africa and meets its domestic energy demand through these resources. Oil based power plants generate 25 TWh of electricity and gas based power plants generate 117 TWh of electricity (Table 2.2). Oil production has declined drastically in the recent past and the current production is only 720,000 bbl/day and unable to meet the current 3% growth in energy demand. A large part of the oil produced is consumed domestically and the country is importing oil (diesel) to meet the demand. In addition to this, subsidies are affecting the country's economy to a large extent because the country's energy- intensive economy is what supports tourism and manufacturing. These two large economic sectors accounts for 25% of the country's GDP. Hence, under the current status Egypt needs an energy source mix to tone up its economy and should develop its geothermal energy sector to save oil and gas for the domestic sector and increase exports to sustain healthy GDP growth (section 2.2.1).

At present the main renewable energy source is wind, which is generating about 550 MWe and expected to grow to 8,760 MWe by 2022 (El Sobki et al., 2009). The New and Renewable Energy Association (NREA) is currently selling wind power to a government controlled transmission company at the rate of 2.5 US cents/kWh (actual cost is 8 to 12 US cents/kWh) which is lower than the generation cost. According to the American Wind Association, the current cost of wind generated power is around 5 to 8 US cents/kWh. But wind energy cannot provide baseload electricity and

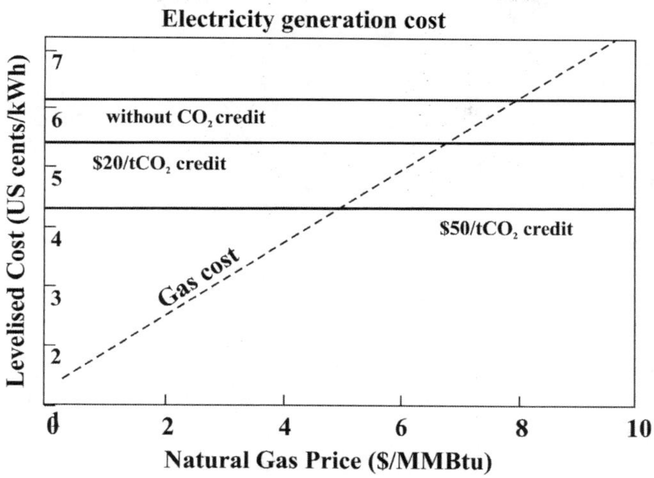

Figure 10.5 Levelised cost of wind power with and without CO_2 subsidy compared to gas power (adapted from El Sobki et al., 2009).

the electricity generated needs to be stored in batteries. This contributes to the additional costs of the tariff. The levelised cost of wind with and without carbon credits compared to the generation cost based on gas sources is shown in Figure 10.5.

At present the carbon credit price is not profitable (as shown in Figure 10.5) and hence the cost of wind energy will be reasonable once the carbon credit price goes beyond \$20/tCO_2. Egypt has alternative sources to wind. The geothermal energy sources discussed at length in Chapter 2 can support an energy mix for the country's energy demand and power generated from geothermal energy can supply baseload power. The levelised cost of geothermal power (hydrothermal) is less than that of wind generated power and geothermal energy can supply baseload with 90% efficiency all 365 days in a year. The life of geothermal power plants is greater than 30 years and most of the power plants in the world work for >60 years (e.g. Larderello power plant in Italy started in 1940). The payback period for geothermal based power projects is around 4 to 5 years. This period is too short for wind as well as solar pv based power projects. If geothermal resources are not available, then countries can depend on wind power as the best option as a power mix source and to obtain carbon credits (e.g. The Netherlands).

The geothermal (hydrothermal) manifestations in Egypt are located around Suez (Chapter 4, section 4.1.2). The estimated geothermal potential of these provinces is 12 to 28 MWe (95 to 221×10^6 kWh; Current electricity generation from oil is 25×10^9 kWh, and total generation is 157×10^9 kWh, see Table 2.2) (Lashin, 2012, Zaher et al., 2012). This amount may be less than 550 MWe generated by wind, but the advantage is that geothermal power can supply baseload power and the cost of electricity is lower than the cost of electricity from wind without subsidy and without any backup power support. Furthermore, Egypt has substantial EGS sources (section 9.3.1) with an average heat generated by granites at 18 μW/m^3. 1 km^3 of high heat generating granite can generate 79×10^6 kWh (Somerville et al., 1994). For example, let us consider the younger high heat generating granitoids exposed El Faliq in the eastern desert. The outcropping area of the granite is 95 km^2. If this granite is developed using EGS technology, then (assuming the granite between 2 and 3 km depth is engineered to create a reservoir), about 7505×10^6 kWh of electricity can be generated. If the entire volume of high heat generating granites from the eastern desert region is exploited, the electricity generated would be substantial. There are two options for the country to increase its GDP: 1) by reducing carbon emissions by substituting part of the electricity demand by geothermal energy or by 2) generating fresh water through desalination using geothermal power. Generating fresh water will save carbon emissions and the fresh water can be used for supporting agricultural activity where Nile water can not be supplied. This would provide the country with food security, energy security and provide fresh water for agricultural and industrial activity at any time irrespective of the weather fluctuations that cause erratic rainfall. Enhancing agricultural activity will increase the gross income and uplift the socio-economic conditions of the rural population. The current CO_2 emission from oil is 90 million tonnes and from gas is 88 million tonnes. Through geothermal energy sources, the country could straight away save 90 million tonnes of CO_2 by offsetting oil. Assuming US\$ 20/tCO_2, the country can save a minimum of 90 million US dollars through carbon credits and earn additional revenue by exporting oil that is saved from domestic use. The fresh water consumption by Egypt is 77 billion m^3. Egypt is facing severe water shortage issues and was seriously considering nuclear energy options to produce fresh water through desalination (Karameld and Mekhermar, 2001). Now that EGS technology is matured, Egypt can exploit all its high heat generating granites for desalination and become a water secured country; this will make the country to have surplus in food as well. Thus geothermal energy can boost the country's GDP through carbon saving, increase in agricultural production and industrial activity.

10.5.2 Saudi Arabia

Saudi Arabia's immediate need is fresh water to support its growing population and agricultural activities. Saudi Arabia's heat generating granites, which occupy an area of 161,467 km^2, are a warehouse of clean energy for the future. Assuming a 2% recovery from such granites, it is estimated, for example, that the Midyan granites alone can generate 160×10^{12} kWh of electricity (Chandrasekharam et al., 2015b). Several such EGS reservoirs are available along the western Arabian shield (see Chapter 9).

The EGS together with the hydrothermal systems can drastically cut CO_2 emissions which are currently exceeding 446,000 Gg from oil and gas based energy sources. Developing these sources will partly offset fossil fuels consumption which is currently generating 240 tera-watt-hours of electricity. What Saudi Arabia can do immediately is to implement GHP technology for space cooling and heating.

Ground thermal conductivity measurements in certain parts of the country have shown that the subsurface ground temperature is suited to install GHP for space cooling in several parts of the country (Chandrasekharam et al., 2014b). Savings from 900 Gg of carbon, which is being sourced from conventional space cooling of commercial and residential establishments, will give additional revenue to the country in the order of 18 million US dollars.

By using geothermal energy as a source mix, the country can: i) reduce its dependence on fossil fuel and decrease the domestic share of the oil there by increasing foreign earnings. This will also help the country to extend the life of the reservoirs, ii) retain supremacy in the world of oil production, iii) rely on indigenous agricultural produce rather than importing staple food grains like wheat and barley, iv) support the production of wheat and barley through irrigation for several decades, and v) have a surplus fresh water resources and support country's demand for fresh water for several decades. Energy optimisation models reveal that by using a source mix (fossil fuel and renewables) the country can reduce the cost of power by 28% per year from 2030 onwards and can save the equivalent of about 0.5 Mb/day of oil 2020 onwards which can be exported (Chandrasekharam et al., 2015, Hussein et al., 2013). Desalination plants benefit most in Saudi Arabia if the geothermal sources are used for this purpose. An economic analysis for a 20,000 m^3/day vacuum membrane distillation (VMD) plant reveals that the production cost of fresh water generated with geothermal energy is US$ 0.53/m^3 while the cost of water generated using a conventional energy source is US$ 1.22/m^3 (Sarbatly and Chiam, 2013).

At present the per capita water demand of 275 L/day is generated using 134×10^6 kWh of electricity and in future this demand will increase by 20%. By offsetting fossil fuel use for desalination processes by using geothermal sources, Saudi Arabia can further save carbon emissions. In addition to drinking water produced through desalination, the country needs water for agricultural activities, especially for producing staple food like wheat and barley and commercial crops like date palms. Due to severe water scarcity, the country has adopted a policy to phase out the production of wheat by 2016. Wheat is an important item in the Saudi diet, consumed commonly in the form of pita. Average per capita consumption is 241 gm per day or 88 kg annually. The total wheat consumption in Saudi Arabia in MY 2013–2014 was about 3.25 million metric tons. Due to the new government policy, wheat imports are expected to reach 3.03 million metric tons in MY 2013–2014 compared to 1.92 million metric tons in 2012–2013 (USDA, 2013). Depending on imported food and outsourcing production of agricultural products are disastrous for any country. Food security should be within the purview of the country. A solution to support agricultural activities exists in the country. The water stress situation related to agriculture can be mitigated if only the country realises this situation and develops its huge untapped geothermal energy resources (Chandrasekharam et al., 2015d). A drastic change in energy and food security policies is essential for the country to support sustained development and maintain supremacy in energy production and supply.

10.5.3 Republic of Yemen

Republic of Yemen is the poorest country located around the Red Sea with a current GDP of 35.6 billion US dollars (see Chapter 2). It is completely dependent on other countries for energy and food.

Currently 721 million US dollars' worth of food is consumed and 416 million US dollars' worth of energy (oil) is imported by the country. Because of the lack of food and energy security, the country's poverty level has increased to 55% in 2009 from 42% in 2004. About 45% of the population starve without food and water (see Chapter 4). This is due to a non-effective energy and food policy and lack of initiative to utilise the renewable energy available within the country. Earlier economic and technical feasibility studies to generate geothermal power at Damt and Damar geothermal provinces have been completed and a detailed project report to generate 1 MWe geothermal power has been approved by the World Bank for financial support in 2009 for an exploratory drill hole. However, the project has still not taken off until now and under the present political situation it may be difficult to restart the project.

10.5.4 Djibouti

Djibouti is the poorest country among the Red Sea countries with its rural population living in utter poverty without access to amenities like power, water and food. Only Djibouti city has a good support system with 100% imported energy and food. The geothermal energy resources are distributed over the entire country, extending from Lake Asal near the Red Sea to Lake Abhe towards the west near the Ethiopian border adjacent to the Damah Ale volcano. The geothermal systems are of high temperature associated with fumaroles (Lake Asal) and two phase (steam and water) systems (Lake Abhe) (see Chapter 4).

Currently Djibouti is generating 730×10^6 kWh of electricity from diesel based generators emitting 539,000 tonnes of CO_2. Although a geothermal development programme has been initiated more than a decade ago by the Iceland geothermal company, due to a financial crisis, the company could not start the project and several pieces of equipment like the steam separator, wells and associated ancillary support equipment were abandoned (Figure 10.6).

However, the country has planned to develop all its geothermal sites, and projected a phased development (Moussa and Suleiman, 2015) as shown in Table 10.6.

According to the project proposal (Moussa and Suleiman, 2015), by the year 2020 the Lake Asal site is scheduled to develop 100 MWe and the rest of the sites will initiate small projects of 1 MWe each between 2015 and 2018. If fully implemented, the projects will to a certain extent, reduce the poverty level by providing water and electricity to the rural areas and lift the socio-economic status of the rural population. As discussed in Chapter 4, the geothermal potential is much greater than 113 MWe and hence the country can generate its own electricity and cultivate food and become energy and food secured. If 113 MWe (891×10^6 kWh) is generated by 2020, then, the country can completely offset fossil fuel and become self-sufficient with respect to energy and food (including water). This will be the only country around the Red Sea that can have an electricity surplus by developing all its geothermal sites. In addition to savings from imported energy and food, CO_2 emissions reduction can further enhance the country's economy making its GDP very strong. In 2009 Djibouti imported fuel and food and other goods worth 647 million US dollars out of which 41 million US dollars was the share of petroleum related products including diesel, 26 million US dollars is the share of food products. In addition to the above share of revenue, carbon savings would earn about 11 million US$ at the rate of 20 US$/t$CO_2$ saved. The Precambrian basement or granites are not present in Djibouti since the country is located within Afar that is floored by oceanic crust covered by Jurassic to Cretaceous sediments and volcanic rocks (see Chapter 4). Hence the country has no EGS source and has to depend only on hydrothermal resources.

Figure 10.6 Abandoned geothermal surface equipment, Lake Asal, Djibouti. (a) and (b) Well heads, (c) Steam separator (Photos by D. Chandrasekharam).

Table 10.6 Geothermal development time table, showing project schedule (Source: Moussa and Suleiman, 2015).

Geothermal site	*Enthalpy*	*Size of project MWe*
L. Asal + Ghoubbet	High	100
L. Abhe	Medium	1
Remaining sites	Medium	12

10.5.5 Eritrea

Among the countries around the Red Sea, next to Djibouti, Eritrea is the least populated but emits CO_2 more or less equal to Djibouti (see Table 2.2 Chapter 2). Its economy runs on imported fuels and food. The total fuel imports amount to 49 million US dollars while its food imports amount to 95 million US dollars. The country uses 0.33 billion kWh of electricity generated from fossil fuel based power plants, emitting 4,921,000 tonnes of CO_2. The advantage of Eritrea is that it has substantial high temperature geothermal systems due to the presence of active volcanoes (Chapter 4). Like Djibouti, the country needs drastic energy policy changes to enhance the use of renewable energy sources like geothermal energy. Due to the current unstable political scenario, it will take time for the country to come out of its present poor socio-economic status. The geothermal sites are conveniently located along the coast and the country can generate substantial fresh water through desalination using geothermal energy. If not full then part development of its geothermal resources will put the country on the path of economic reform.

References

Abdalla, F.A. and Scheytt, T. 2012. Hydrochemistry of surface water and groundwater from a fractured carbonate aquifer in the Helwan area, Egypt. *J. Earth Syst. Sci.* 121, 109–124.

Abdallah, S. A. E., Mohamed, G.A. and Bakry, A.R. 2014. Rare metal mineralization (Zr, U, Th, REE) associated to El Seboah acidic peralkaline rocks, south-western desert of Egypt: Recovery technique. *American J. Earth Sci.* 1, 73–85.

ADB, 2013. African Development Bank group: Geothermal Exploration Project in the Lake Assal region: Djibouti. Project appraisal report. 21p.

Agar, R.A. 1992. The tectono-metallogenic evolution of the Arabian shield. *Precambrian Res.* 58, 169–194.

Ahmed, A. H. and Habtoor, A. 2015. Heterogeneously depleted Precambrian lithosphere from mantle peridotites and associated chromite deposits of Al'Ays ophiolite, Northwestern Arabian Shield. *Ore Geol.* 67, 279–296.

Ahmad, A. and Ramana, M.V. 2014. Too costly to matter: Economics of nuclear power for Saudi Arabia. *Energy*. 69, 682–694.

Al-Amri, A. M. S. 1994. Seismicity of the south-western region of the Arabian Shield and southern southern Red Sea. *J. African Earth Sci.* 19, 17–25.

Al-Kadasi, M., Menzies, M., Baker, J., 1999. K–Ar chronology of the basalt-rhyolite province in western Yemen: the need for data screening. *Yemeni J. Sci.* 1, 67–80.

Almazroui, M., Islam, M.N., Athar, H., Jones, P.D. and Rahman, M.A. 2012. Recent climate change in the Arabian peninsula: annual rainfall and temperature analysis of Saudi Arabia. *Inter. J. Climatology*, 32, 953–966.

Angelier, J. 1985. Extension and rifting: the Zeit region, Gulf of Suez. *J. Structural Geol.* 1, 605–612.

Al Saleh, A.M. and Kassem, M.K. 2012. Microstructural finite strain analysis and 40Ar/39Ar evidence for the origin of the Mizil gneiss dome, eastern Arabian Shield, Saudi Arabia. *J. African Earth Sci.* 70, 24–35.

Al Shanti, A.M. and Roobol, M.J. 1979. Some thoughts on metallogenesis and evolution of the Arabian-Nubian shield in Evolution and mineralization of the Arabian Nubian Shield-Proceedings, (Ed) A.M.S. Al Shanti, Pergamon Press, London, 87–96.

Al-Shanti, A.M.S and Mitchell, A.H.G. 1976. Late Precambrian subduction and collision in the Al Amar-Idsas region, Arabian Shield, Kingdom of Saudi Arabia. *Tectonophysics*, 30, T41–T47.

Asif, M. and Alrashed, F. 2012. Prospects of renewable energy to promote zero energy residential buildings in the KSA. *Energy Procedia*. 18, 1096–1105.

As-Saruri, M.L., 1999. Lithostratigraphie der Tertiar-Sedimente der Republik Jemen. Schriftenreihe fur geologische Wissenschaften H. 10, Berlin, Federal Republic of Germany.

Banat, K.M., Howari, F.M. and Kadi, K.A. 2005. Water Chemical Characteristics of the Red Sea Coastal Sabkhas and Associate Evaporite and Carbonate Minerals. *J. Coasl Res.*, 21, 1068–1081.

Baria, R., Baumgärtner, J., Gérard, A. and Garnish, J. 2000. The European HDR programme: main targets and results of the deepening of the well GPK2 to 5000. *Proceedings of the World Geothermal Congress, Kyushu – Tohoku, Japan.*

Baria, R., Michelet, S., Baumgärtner, J., Dyer, B., Gerard, A., Nicholls, J., Hettkamp, T., Teza, D., Soma, N. and Asanuma, H., 2004. Microseismic monitoring of the world largest potential HDR reservoir. *Proceedings of the 29th Workshop on Geothermal Reservoir Engineering, Stanford University, California.*

Baria, R., Baumgartner, J., Gérard, A. and Jung, R. 1998. "European Hot Dry Rock geothermal research programme 1996–1997." Contract N°: JOR3CT950054, Joule III Programme, final report EUR 18925 EN, 151pp.

Baria, R., Jung, R., Tischner, T., Nicholls, J., Michelet, S., Sanjuan, B., Soma, N., Asanuma, H., Dyer, B. and Garnish, J. 2006. Creation of an HDR reservoir at 5000 M depth at the European HDR project. Proceedings, Thirty-First Workshop on Geothermal Reservoir Engineering Stanford University, Stanford, California, January 30–February 1, 2006 SGP-TR-179.

Batchelor, A. S. 1982. "The stimulation of a Hot Dry Rock geothermal reservoir in the Cornubian Granite, England." *Proc. 8Workshop on Geothermal Reservoir Engineering*, Stanford, Ca, USA, 14–16 Dec. pp. 237–248.

Battistellia, A., Yiheyis, A., Calore, C., Ferragina, C. and Abatneh, W. 2002. Reservoir engineering assessment of Dubti geothermal field, Northern Tendaho Rift, Ethiopia. *Geothermics* 31, 381–406.

BP, 2013. BP (British Petroleum) Statistical Review of World Energy June 2013. 48p

Bakor, A.R, Gass, I.G., Neary, C.R. 1976. Jabal Wask, NW Saudi Arabia, An Eocambrian back arc ophiolite. *Earth. Planet. Sci. Letters* 30, 1–9.

Bayer, H.J., El Isa, Z., Hotzl, H., Mechie, J., Prodehl, C. and Saffarini, G. 1989. Large tectonic and lithospheric structures of the Red Sea region tectonic and lithospheric structures of the Red Sea region. *J. African Earth Sci.* 8, (2/3/4), 565–587.

Barberi, F., Ferrara, R. Santacroce, R. and Varet, J. 1975. Structural evolution of the Afar triple junction, in Afar depression of Ethiopia, edited by A. Pilger and A. Roesler, pp. 39–53, Schweizerbart, Stuttgart, Germany.

Bartle, A. 2002. Hydropower potential and development activities. *Energy Policy*, 30, 121–1239.

Bentor, Y.K. 1985. The crustal evolution of the Arabo-Nubian Massif with special reference to the Sinai Peninsula. Precambrian Research 28, 1–74.

Bokhari, M.M., Jackson, N. and Al-Oweidi K. 1986. Geology and mineralization of the Jabal Umm Al Suqian albitized apogranite, south Najd region, Kingdom of Saudi Arabia. *J. African Earth Sci.* 4, 189–198.

Bregar, M., Bauernhofer, A., Pelz, K., Kloetzli, U., Fritz, H., Neumayr, P., 2002. A late Neoproterozoic magmatic core complex in the Eastern Desert of Egypt: emplacement of granitoids in a wrench-tectonic setting. Precambrian Research 118, 59–82.

Beydoun, Z.R., As-Saruri, A.L., Mustafa, El-Nakhal, H., Al-Ganad, I.N., Baraba, R.S., Nani, A.S.O. and Al-Aawah, M.H., 1998. International lexicon of stratigraphy v.III, Republic of Yemen. In: 2nd ed.: International Union of Geological Sciences and Ministry of Oil and Mineral Resources, Republic of Yemen, Publication 34.

Black, M., Morton, W.H. and Rex, D.C. 1974. "Block Tilting and Volcanism within the Afar in the Light of Recent K/Ar Age Data." In *Afar Depression of Ethiopia*, edited by A. Pilger and A. Rösler, 296–299. Germany: Bad Bergzarben, F.R.

Bohannon, R.G. 1986. How much divergence has occurred between Africa and Arabia as a result of the opening of the Red Sea? *Geology*, 14, 510–513.

Bohannon, R.G., Naeser, C.W. and Schmidt, L.S. 1989. Zimmermann RA. The Timing of Uplift, Volcanism, and Rifting Peripheral to the Red Sea: A Case for Passive Rifting? *J. Geophy. Res.* 94, 1683–1701.

Bond, T., Bhardwaj, E., Dong, R., Jogani, R., Jung, S., Roden, C., Streets, D.G. and Trautmann, N.M. 2007. Historical emissions of black carbon and organic carbon aerosols from energy related combustion, 1850–2000. *Global Biogeochemical Cyc.*, 21, doi:10.1029/2006GB002840, 2007.

Bosch, B., Deschamps, J., Leleu, M., Lopoukhine, M., Marce, A. and Vilbert, C. 1977. The geothermal zone of lake Assal (FTAI), geochemical and experimental studies. *Geothermics*, 5, 165–175.

Bosworth, W. and Tavian, M. 1996. Late quaternary reorientation of stress field and extension direction in the southern Gulf of Suez, Egypt: evidence from uplifted coral terraces, mesoscopic fault arrays and borehole breakouts. *Tectonics*, 15, 791–802.

Bosworth, W. and McClay, K. 2001. Structural and stratigraphic evolution of the Gulf of Suez Rift, Egypt: a synthesis. *In:* P.A. Ziegler, W. Cavazza, A.H.F. Robertson and S. Crasquin-Soieau (eds), Peri-Tethys Memoir 6: Peri-Tethyan Rift/Wrench Basins and Passive Margins. *Mem Mus. nam. Hist. nat.*, 186:567–606. Paris ISBN:2-85653-528-3.

Bosworth, W., Huchon, P. and McClay, K. 2005. The Red Sea and Gulf of Aden Basins. *J. African Earth Sci.*, 43, 334–378.

Brown, G.F. 1970. Eastern margin of the Red Sea and the coastal structures in Saudi Arabia. *Phil. Trans. Roy. Soc. Lond. A.* 267, 75–87.

Brown, D.W., Duchane, D.V., Heiken, G. and Thomas, V.T. 2012. mining the Earth's heat: Hot dry rock geothermal energy. Springer-Verlag Berlin Heidelberg, 669p.

Buskirk, R.V. 2006. Analysis of long-range clean energy investment scenarios for Eritrea, East Africa. *Energy Policy*, 34.

Camp, V.E. and Roobol, M.J. (1992) Upwelling Asthenosphere Beneath Western Arabia and Its Regional Implications. *J. Geophy. Res.* 97, 15255–15271.

Cataldi, R., Hodgson, S.F. and Lund, J.W. 1999. Stories from a heated earth, our geothermal heritage. *Geothermal Res. Council.* 588p.

Cermak, V., Huckenholz, H.G., Rybach, L., and Schmid, R. 1982. Radioactive heat generation in rocks. In: Hellwege, K. (Ed.), Landolt-Bornstein numerical data and functional relationships in science and technology. New Series, Group V. Geophysics and Space Research, vol. 1, Physical properties of rocks, subvolume b. Springer, Berlin, Heidelberg, New York. 433–481.

Chandrasekhar, V. and Chandraseklharam, D. 2008. Enhanced geothermal resources in NE Deccan Province, India 2008. *Geothermal Res. Council Trans*, 32, 71–75.

Chandrasekharam, D. and Antu, M.C. 1995. Geochemistry of Tattapani thermal springs, Madhya Pradesh, India: Field and experimental investigations. *Geothermics* 24, 553–559.

Chandrasekharam, D. and Prasad, S.R. 1998. Geothermal system in Tapi rift basin, northern Deccan province, India. In: G.B. Arehart and J.R. Hulston (eds): *Proceedings 9th Water-Rock Interaction*. A.A. Balkema, Leiden, The Netherlands, 667–670.

Chandrasekharam, D. 2001. Use of Geothermal energy for food processing: Indian Status. Quart. Bull., *Geo-heat Centre*, Oregon, U.S.A. 22, 8–12.

Chandrasekharam, D. and Chandrasekhar, V. 2008. Granites and granites: India's warehouse of EGS, Bull. *Geother. Res. Council*, 37, 17–20.

Chandrasekhar, V. and Chandrasekharam, D. 2009. Geothermal Systems in India. Geothermal. Res. Council Trans, 33, 607–610.

Chandrasekharam, D. and Bundschuh, J. 2008. "*Low Enthalpy Geothermal Resources for Power generation" Taylor and Francis Pub., U.K. 169 pp.*

Chandrasekharam, D. and Chandrasekhar, V. 2010a. "Geothermal Energy Development of L. Abbe and CDM for Djibouti." *Proceedings. ArGeo 3 Conference, Djibouti.*

Chandrasekharam, D. and Chandrasekhar, V. 2010b. Hot Dry Rock Potential in India: Future Road Map to Make India Energy Independent. *Proceed. World Geothermal Congress 2010, Bali, Indonesia.*

Chandarasekharam, D., Lashin, A. and Al Arifi, N. 2014a. CO_2 mitigation strategy through geothermal energy, Saudi Arabia. *Renew. Sustain. Energy Rev.* 38, 154–163.

Chandarasekharam, D., Lashin, A. and Al Arifi, N. 2014b. The potential contribution of geothermal energy to electricity supply in Saudi Arabia. *Inter. J. Sustainable Energy.* http://dx.doi.org/10.1080/14786451.2014.950966.

Chandrasekharam, D., Lashin, A., Al Arifi, N., Chandrasekhar, V. and Al Bassam, A. 2015a. Clean Development Mechanism through Geothermal, Saudi Arabia. *Proceed. World Geothermal Congress*, 2015, Melbourne, Australia, April 2015.

Chandrasekharam, D., Lashin, A., Al Arifi, N., Al Bassam, A., Ranjith, P. G., Varun, C. and Singh, H.K. 2015b. Geothermal energy resources of Jizan, SW Saudi Arabia *J. African Earth Sci.* 109, 55–67.

Chandrasekharam, D., Lashin, A., Al Arifi, N., Al Bassam, A. and Varun, C. 2015c. Evolution of geothermal systems around Red Sea. *Environ. Earth. Sci.* 73, 4215–4236.

Chandrasekharam, D., Lashin, A., Al Arifi, N., Al Bassam, A., El Alfy, M., Ranjith, P. G., Varun, C. and Singh, H.K. 2015d. CO_2 emission and climate change mitigation using the enhanced geothermal system (EGS) based on the high radiogenic granites of the western Saudi Arabian shield. *J African Earth Sci.* 112, 213–233.

Chaudhuri, H., Sinha, B. and Chandrasekharam, D. 2015. Helium from geothermal sources. *Proceed. World Geothermal Congress, 2015, Melbourne, Australia*, 19–25 April 2015.

Chierici, L. 1964. Planning of geothermo-electric power plant: technical and economic principles. In Geothermal Energy II, UN Conference on New Sources of Energy, Rome, 1961, v3, 299–311.

Chowdhury S. and Al Zahrani, M. 2015. Characterizing water resources and trends of sector wise water consumption in Saudi Arabia. *J King Saud Uni: Engineering Scis.* 27, 68–82.

CNRS-CNR (Centre National de la Recherche Scientifique-Consiglio Nazionale delle Ricerche). 1973. "Geology of Northern Afar (Ethiopia)." *Rev. Géog. Phys. Géol. Dyn. (2)* XV (4): 443–490.

Chessex, R., M. Delaloye, J. Muller, and M. Weidmann, Evolution of the volcanic region of Ali Sabieh (T. F. A. I.), in the light of K-Ar age determinations, in Afar Depression of Ethiopia, vol. 1, edited by A. Pilger and A. Röesler, pp. 221–227, Schweizerbart, Stuttgart, Germany, 1975.

Cochran, J.R. and Martinez, F. 1988. Evidence from the northern Red Sea on the transition from continental to oceanic rifting. *Tectonophy.* 153, 25–53.

Colletta, B., Le Quellec, P., Letouzey, J. and Moretti, I. 1988. Longitudinal evolution of the Suez rift structure (Egypt). *Tectonophy.* 153, 221–233.

Coleman, R.G., Gregory, R.T and Brown, G.F. 1983. Cenozoic volcanic rocks of Saudi Arabia. U S G S Open file report, 83–788.

Cooper, M.A., Herbert, R. and Hill, G.S. 1989. The structural evolution of Triassic intermontane basins in northeastern Thailand. *Proceedings International Symposium on Intermonte Basins: Geology, and Resources*, Chiang Mai, Thailand. 231–242.

Coulié, E., Quidelleur, X., Gillot, P.-Y., Courtillot, V., Lefévre, J.-C. and Chiesa, S. 2003. Comparative K-Ar and Ar/Ar dating of Ethiopian and Yemenite Oligocene volcanism: implications for timing and duration of the Ethiopian traps. *Earth and Planetary Science Letters* 206, 477–492.

Craig, H. 1961. Standard for reporting concentrations of deuterium and oxygen-18 in natural waters. *Science* 133, 1833–1934.

D'Almeida, G.A.F. 2010. Structural Evolution History of the Red Sea Rift. *Geotectonics*, 44, 271–282.

Dawood, Y.H., Harbi, H.M. and Abd El-Naby, H.H. 2010. Genesis of kasolite associated with aplite-pegmatite at JabalSayid, Hijaz region, Kingdon1 of Saudi Arabia. *J. Asian Earth Sci.* 37, 1–9.

Debayle, E., Leveque, J.J. and Cara, M. 2001. Seismic evidence for a deeply rooted low velocity anomaly in the upper mantle beneath the north-eastern Afro/Arabian continent. *Earth. Planet Sci. Lett.* 193, 423–436.

Deniel, C., Vidal, P., Coulon, C., Vellutini, P.J. and Piguet, P. 1994. Temporal evolution of mantle sources during continental rifting: The volcanism of Djibouti (Afar), *J. Geophys. Res.*, 99(B2), 2853–2870.

Debayle, E., Leveque, J.J. and Cara, M. 2001. Seismic evidence for a deeply rooted low-velocity anomaly in the upper mantle beneath the north eastern Afro/Arabian continent. *Earth. Planet. Sci. Lett.* 193, 423–436.

de la Vega, F.F. 2010. Impact of Energy Demand on Egypt's Oil and Natural Gas Reserves: Current situation and perspectives to 2030 'in' J. Sawin and L. Mastny (Eds), "Egyptian-German Joint Committee on Renewable Energy, Energy Efficiency and Environmental Protection", Pub: Deutsche Gesellschaft für Technische Zusammenarbeit (GTZ) GmbH, 1–83.

Di Pippo, R. 2008. Geothermal power plants: Principles, applications, Case studies and environmental impact. Butterworth-Heinemann is an imprint of Elsevier, Amsterdam, 587p.

Dor Ji and Ping, Z.N. 2000. Characteristics and genesis of the Yangbajing geothermal field, Tibet. *Proceedings World Geothermal Congress 2000*, May 28–June 10, Kyushu-Tohoku, Japan, pp. 1083–1088.

Diao, X. and Pratt, A.N. 2007. "Growth options and poverty reduction in Ethiopia – An economy-wide model analysis." *Food Policy* 32, 205–228.

Didana, Y.L., Thiel, S. and Heinson, G. 2015. Magnetotelluric Imaging at Tendaho High Temperature Geothermal Field in North East Ethiopia. *Proceed. World Geothermal Cong., 2015*, Melbourne, Australia.

Dercon, S., Hoddinott, J. and Woldehanna, T. 2005. Shocks and Consumption in 15 Ethiopian Villages, 1999–2004. *J. African Economies* doi:10.1093/jae/eji022.

Duncan, R.A. and Al Amri, M. 2013. Timing and composition of volcanic activity at Harrat Lunayyir, western Saudi Arabia. *J. Volcanol. Geother. Res.*, 260, 103–116.

Duffield., W.A., Bullen, T.D. Clynne, M.A. Fournier, R.O., Janik, C.J., Lanphere, M.A., Lowenstern, J., Smith, J.G., Giorgis, L., Kahsai, G., Mariam, K. and Tesfai, T. 1997. Geothermal potential of the Alid volcanic center, Danakil Depression, Eritrea. US Geological Survey Open File Report, 291, 62 p.

du Bray E. 1986. Jabai Silsilah tin prospect, Najd region, Kingdom of Saudi Arabia. *J. African Earth Sci.* 4, 237–247.

du Bray E.A., Elliott, J.E. and Stoeser, D.B. 1983. Geochemical evaluation of felsic plutonic rocks in the eastern and southeastern Arabian shield, Kingdom of Saudi Arabia. USGS Open file report 83-369.

Eby, G.N., 1992. Chemical subdivision of the A-type granitoids; petrogenetic and tectonic implications. *Geology* 20, 641–644.

El Ramly, M.F. and Akaad, M.K. 1960. The basement complex in the central-eastern desert of Egypt between latitudes 24°30′ and 25°40′N. *Geological Survey of Egypt* 8, 1–35.

Ellis, A.J. and Mahon, W.A.J. 1967. Natural hydrothermal systems and experimental hot water/rock interactions (Part II). *Geochim. Cosmochim. Acta,* 31, 519–538.

Ellis, A.J. and Mahon, W.A. 1977. *Chemistry and geothermal systems.* Academy Press, New York, NY.

Eltamaly, A.M. 2013. Design and implementation of wind energy system in Saudi Arabia. *Renew. Energy*, 60, 42–52.

El Ahmady Ibrahim, M., El Hamed El Kalioby, B.A., Aly, M.G., El Tohamy A.M. and Watanabe, K. 2015. Altered granitic rocks, Nusab El Balgum Area, Southwestern Desert, Egypt: Mineralogical and geochemical aspects of REEs. *Ore Geology Rev.* 70, 252–261.

El Khashab, H. and Al Ghamedi, M. 2015. Comparison between hybrid renewable energy systems in Saudi Arabia. *J. Electrical Sys. Inform. Tech.* (in press).

El Juhany, L.I. 2008. Degradation of date palm trees and data production in Arab countries: Causes and potential rehabilitation. *Australian J Basic and App. Sci.*, 4, 3998–4010.

El Sobki, M., Wooders, P. and Sherif, Y. 2009. Clean energy investment in developing countries: wind power in Egypt. International Institute for Sustainable Development (www.iisd.org), Trade, Investment and climate change series report, 54p.

Elliott, J. E. 1983. Peralkaline and peralumnois granites and related mineral deposits of the Arabian Shield, Kingdom of Saudi Arabia. USGS Open file report 1983; 83–389:37.

Elsayed, R.A.M., Assran, H.M. and Elatta, S.A.A. 2014. Petrographic, radiometric and paleomagnetic studies for some alkaline rocks, south Nusab El Balgum mass complex, south western Egypt. *Geomaterials*, 4, 27–46.

EIA, 2013. Annual Energy Outlook 2013 with Projection to 2040. US Energy Information Administration, 244p.

EIA, 2015. Annual Energy Outlook 2015. US Energy Information Administration, Wshington, 154p.

Emam, A., Moghazy, N.M., El-Sherif, A.M., 2011: Geochemistry, petrogenesis and radioactivity of El Hudi I-type younger granites, South Eastern Desert, Egypt. *Arab J Geosci*, 4: 863–878.

Evans, T.R. and Tammemagi, H.Y. 1974. Heat flowand heat production in north-east Africa. *Earth Planett. Sci. Lett.* 23, 349–356.

Fara, M., Chandrasekharam, D. and Minissale, A. 1999. Hydrogeochemistry of Damt thermal springs, Yemen Republic. *Geothermics* 28, 241–252.

Falkenmark, M., Lundqvist, J. and Widstrand, C. 1989. Macro-scale water scarcity requires micro scale approaches. Aspects of vulnerability in semi arid development. *Natural Resources Forum*. 13, 258–267.

Faure, G. 1986. Principles of isotope geology. John Wiely and Sons, New York, NY, 1986.

FDRE, 2011. Ethiopia's climate Resilient Green Economy, Green Economy Strategy. Federal Democratic Republic of Ethiopia report, 200p.

FDRE, 2014. Federal Democratic Republic of Ethiopia: Ministry of Finance and Economic Development. *Growth and Transformation Plan*, Annual Report, 2012/13, 119p.

Féraud, G., Zumbo, V., Sebai, A. and Bertrand, H. 1991. 40Ar/39Ar age and duration of tholeiitic magmatism related to the early opening of the Red Sea rift. *Geophysical Research Letters* 18, 195–198.

Fritz, H., Wallbrecher, E., Khudeir, A.A., Abu El Ela, F. and Dallmeyer, D.R. 1996. Formation of Neoproterozoic metamorphic core complexes during oblique convergence (Eastern Desert, Egypt). *Journal of African Earth Sciences* 23, 311–329.

Fritz, H., Loizenbauer, J. and Wallbrecher, E. 2014. Magmatic and solid state structures of the Abu Ziran pluton: deciphering transition from thrusting to extension in the Eastern Desert of Egypt. *Jour. African Earth Sci.*, 99, 122–135.

Fouillac, A.M., Fouillac, C., Cesbron, F., Pillard, F. and Legendre, O. 1989. Water rock interaction between basalt and high salinity fluids in the Asal Rift, Republic of Djibouti. *Chem Geol.*, 76, 271–289.

Fournier, R.O. and Rowe, J.J. 1966. Estimation of underground temperatures from silica content of water from hot springs and wet-steam wells. *American Journal of Science* 264, 685–697.

Fournier, R.O. and Truesdell, A.H. 1973. An empirical Na-K-Ca geothermometer for natural waters. *Geochim. Cosmochim. Acta*, 37, 1255–1275.

Fournier, R.O. and Potter, R.W. 1982. An equation correlating the solubility of quartz in water from 25°C to 900°C at pressure up to 10,000 bars. *Geochimica et Cosmochimca Acta* 46, 1969–1974.

Fournier, R.O. 1983. A method for calculating quartz solubilities in aqueous sodium chloride solutions. *Geochimica et Cosmochimica Acta* 47, 579–586.

Fournier, R.O. 1985. The behavior of silica in hydrothermal solutions. In: B.R. Berger and P.M. Bethke (eds): *Reviews in economic geology, volume 2*. The Economic Geology Publishing Company, Littleton, CO, pp. 45–61.

Fournier, R.O. 1991. Water geothermometers applied to geothermal energy in "Application of geochemistry in geothermal reservoir development" (Ed) Franco D' Amore, Series of technical Guides on the use of geothermal energy, UNITAR/UNDP Centre on Small Energy Resources. 411p.

Gaafar, I. 2014. Geophysical mapping, geochemical evidence and mineralogy for Nuweibi Rare Metal Albite granite, Eastern Desert, Egypt. *Open Journal of Geology*, 4, 108–136.

Garson, M.S. and Krs, M. 1976. Geophysical and geological evidence of the relationship of Red Sea transverse tectonics to ancient fractures. *Bull. Geol. Soc. Am.* 87, 169–181.

Gaulier, J.M., Le Pichon, X., Lyberjs, N., Avedjk, F., Geli, L., Moretti, I., Deschamps, A. and Hafez, S. 1988. Seismic study of the crust of the northern Red Sea and Gulf of Suez. *Tectonophysics*, 153, 55–88.

Gass, J.G., 1981. Pan-African (Upper Proterozoic) plate tectonic of the Arabian-Nubian Shield. In: Kröner, A. (Ed.), *Precambrian Plate Tectonics*, pp. 387–403.

Gasse, F., Richard, O., Robe, D. and Et Williams, M.A.J. 1980. "Evolution Tectonique et climatique de l'Afar Central d'après les sédiments plio-pléistocènes." *Bulletin de la Societe Géologique France* 7, XXII (6): 987–1001.

Gat, J.R. 1966. Oxygen and hydrogen isotopes in the hydrologic cycle. Annul. Rev. Earth. Planet. Sci. 24, 225–262.

Gaulier, J.M. and Huchon, P. 1990. "Evolution tectonique de l'Afar méridional depuis 3, 5 Ma." Sciences et Techniques no 3, ISERST, Djibouti, 7–12.

Gawthorpe, R. L., Sharp, T., Underhill, J.R. and Gupta, S. 1997. Linked sequence Stratigraphy and structural evolution of propagating normal faults. *Geology*, 25, 795–798.

Gettings, M.E. 1982. Heat flow measurements at shot points along the 1978 Saudi Arabian seismic deep refraction line, Part 2: Discussion and interpretation. USGS Open File report 82-794.

Gettings, M.E., Blank, H.R., Mooney, W.D. and Healey, J.H. 1986. Crustal Structure of South-western Saudi Arabia. *J. Geophy. Res.*, 91, 6491–6512.

Gettings, M. E. and Showail, A. 1982. Heat flow measurements at shot points along the 197X Saudi Arabian seismic deep-refraction line, part I: results of the measurements, U.S. Geol. Surv. Open File Rept. 82-793, 98 pp.

Giggenbach, W.F., Gonfiantini, R. and Panichi, C. 1983. Geothermal systems. In: *Guide book on nuclear techniques in hydrology*. IAEA, Vienna, Tech. Rep. 91, 359–379.

Girdler, R.W. 1970. A review of Red Sea heat flow. Phil. Trans. Roy. Soc. London A, 267, 191–203.

Girdler, R. W. 1970. A review of Red Sea heat flow. Phil. Trans. Roy. So. Land. A. 267: 191–203.

Girdler, R. W. and Evans, T. R. 1977. Red Sea heat flow, Geophy. J. Roy. Astra. So., 51: 245–251.

Giggenbach, W.F. 1988. Geothermal solute equilibria. Derivation of Na-K-Mg-Ca geoindicators. *Geochim. Cosmochim. Acta*. 52, 2749–2765.

Giggenbach, W. 1992. Isotopic shifts in waters from geothermal and volcanic systems along convergent plate boundaries and their origin. *Earth. Planet. Sci. Lett.*, 113, 495–510.

Giggenbach, W. 1998. The isotopic composition of waters from the El Tatio geothermal field, Northern Chile. *Geochim. Cosmochim. Acta.*, 42, 979–988.

Giraud, A., Thouvenot, F. and Huber, R. 1986. Tectonic stress in the southwestern Arabian Shield. *Engg. Geo.* 22, 247–255.

Girdler, R. W. 1966. The role of translational and rotational movement in the formation of the Red Sea and Gulf of Aden: Proceed. Sym. World Rift Systems, Ottawa 1965. Geol. Sur. Canada paper. 66-14, 65–77.

Greene, D.C. 1984. Structural geology of the Qusier area, Red Sea cost, Egypt. Dept. Geology and Geography contribution number 52, University of Massachusetts, USA, 179 p.

Georgsson, L.S., Saemundsson, K. and Hjartarson, H. 2005. Exploration and development of the Hveravellir geothermal field, N-Iceland. In: *Proceedings, World Geothermal Congress 2005*, Antalya, Turkey, p. 10.

Gundmundsson, J.S. and Lund, J.W. 1987. Direct use of earth's heat. In "Applied Geothermics" (Eds) M.J. Economides and P.O. Ungemach. John Wiley & Sons, 189–219.

Hamimi, Z., El Shafei, M., Kattu, G. and Matsah, M. 2013. Transpressional regime in southern Arabian Shield: Insights from WadiYiba area, Saudi Arabia. *Miner. Petrol.* 107, 849–860.

Hamimi, Z., El Sawy, K., El Fakharani, A., Matash, M., Shujoon, A. and El Shafei, M.K. 2014. Neoproterozoic structural evolution of the NE trending Ad-Damm shear zone, Arabian shield, Saudi Arabia. *J. African Earth Sci.*, 99, 51–63.

Hassan, M.A. and Hashad, A.H. 1990. Precambrian of Egypt. In: Said, R. (Ed.), *The Geology of Egypt*. Balkema, Rotterdam, pp. 201–248.

Hamlyn, J.E., Keir, D., Wright, T.J., Neuberg, J.W., Goitom, B., Hammond, J.O.S. Pagli, C., Oppenheimer, C. Kendall, J.-M. and Grandin, R. 2014. Seismicity and subsidence following the 2011 Nabro eruption, Eritrea: Insights into the plumbing system of an off-rift volcano, *J. Geophys. Res. Solid Earth*, 119, 8267–8282, doi:10.1002/2014JB011395

Harrison, T.M., Groove, M., Lovera, O.M. and Catlos, E.J. 1998. A model for the origin of Himalayan antexis and inverted metamorphism. *J. Geophysical Research* 103, 27,017–27,032.

Harrison, T.M., Grove, M., McKeegan, K.D., Coath, C.D., Lovera, O.M. and Le Foort, P. 1999. Origin and episodic emplacement of the Manaslu intrusive complex, Central Himalayas. *J. Petrology* 40, 3–19.

Harrison, T.M., Groove, M., Lovera, O.M. and Catlos, E.J. 1998. A model for the origin of Himalayan antexis and inverted metamorphism. *J. Geophysical Research* 103, 27,017–27,032.

Harrison, T.M., Grove, M., McKeegan, K.D., Coath, C.D., Lovera, O.M. and Le Foort, P. 1999. Origin and episodic emplacement of the Manaslu intrusive complex, Central Himalayas. *J. Petrology* 40, 3–19.

Hance, C.N. 2005. Factors Affecting Costs of Geothermal Power Development. Publication by the Geothermal Energy Association for the US Department of Energy. 64p.

Hirschberg, S., Wiemer, S. and Burgherr, P. 2015. Energy from the Earth: Deep geothermal as a resource for the future? Die Deutsche Nationalbibliothek, Zurich. 526p.

Hersir, G.P. and Flovenz, O.G. 2013. Resistivity surveying and electromagnetic methods. International Geothermal Association (IGA) Report 0110-2013, 10p.

Hettiarachchi, H.M.D., Golubovic, M., Woreki, W.M. and Ikegami, Y. 2007. The performance of the Kalina cycle system11, KCS-11. With Low temperature heat sources. *J. Energy Resources Technol.*, 129, 243–247.

Hoffmann, C., Courtillot, V., Féraud, G., Rochette, P., Yirgu, G., Ketefo, E. and Pik, R. 1997. Timing of the Ethiopian flood basalt event and implications for plume birth and global change. *Nature* 389, 838–841.

Hogarth, R., Holl, H. and McMahon, A. 2013. Flow Testing Results from Habanero EGS Project. *Proceed. Sixth Annual Australian Geothermal Energy Conference*, Nov. 14–15, 2013.

Hori, Y., Kitano, K., Kaieda, H. and Kihl, K. 1999. Present status of the Ogachi HDR project, Japan and future plans. *Geothermics*, 28, 637–645.

Hussein, M.T., Loni, O.A. 2011. Major ionic composition of Jizan thermal springs, Saudi Arabia. *J. Emerging Trends in Eng. Applied Sci.* 2, 190–196

Husseini, M. J. 1988. The Arabian infra Cambrian extensional system. *Tectonophy* 1988; 148:93–103.

Houssein, B., Chandrasekharam, D., Varun, C. and Jalludin, M. 2013. Geochemistry of thermal springs around Lake Abhe, Western Djibouti. *J. Sustainable Energy*, dx.doi.org/10.1080/14786451.2013.813027.

ICPAC (IGAD Climate Prediction and Applications Centre). 2007. "Climate change and human development in Africa: Assessing the risks and vulnerability of climate change in Kenya, Malawi and Ethiopia." United Nations Development Programme, 215p.

IAEA 2005. Isotopic composition of precipitation in the Mediterranean basin in relation to air circulation pattern and climate, Final project Report, IAEA TECDOC 1453, 230p.

Ibrahim, M.E., El Kalioby, B.A.E., Aly, G.M., El Tohamy, A.M. and Watanabe, K. 2015. Altered granitic rocks, Nusab El Balgum Area, Southwestern Desert, Egypt: Mineralogical and geochemical aspects of REEs. *Ore Geology Reviews* 70 (2015) 252–261.

IEA, 2012. International Energy Agency: World Energy Outlook, 134 p.

IEA, 2013. International Energy Agency, Redrawing the energy-climate map, World Energy Outlook special report. 134 p.

IEA, 2014a. Oil market report. International Energy Association. 64p.

IEA. 2014b. CO_2 emissions from fuel combustion: highlights. International Energy Agency report, 136p.

IEA. 2014c. Africa Energy Outlook. International Energy agency: World Energy Outlook special report, 242p.

IEA, 2014d. The way forward: Five key actions to achieve a low carbon energy sector. Report, 8p

IEA. 2014e. Energy technology perspective 2014: harnessing electricity's potential, 382 p.

IRENA. 2014. http://www.irena.org/REmaps/countryprofiles/africa/Ethiopia.pdf#zoom=75 (accessed on 18 Sept. 2014).

Johnson, P.R. and Woldehaimanot, B. 2003. Development of the Arabian-Nubian Shield: perspective on accretion and deformation in the northern East African Orogen and the assembly of Gondwana, in "Proterozoic East Gondwana: Supercontinent Assembly and Breakup" (Eds). M. Yoshida, B.F. Windley and S. Dasgupta, *Geol. Soci. London Spl. Pub.* 206, 289–325.

James, R. 1968. Wairakei and Larderello geothermal systems compared. *New Zealand J. Sci and Tech.* 11, 706–719.

Jonsson, M.: *Advanced power cycles with mixtures as the working fluid*. PhD Thesis, Royal Institute of Technology, Stockholm, Sweden, 2003.

Jarrige, J.J., Ott D'Estevou, P., Burollet, P.F., Montenat, C., Prat, P., Richert, J.P. and Thiriet, J.P. 1990. The multistage tectonic evolution of the Gulf of Suez and northern Red Sea continental rift from field observations. *Tectonics*, 9, 441–465.

Kaila, K.L., Rao, I.B.P., Koteswar Rao, P., Madhava Rao, Krishna, V.G. and Sridhar, A.R. 1981. DSS studies over Deccan trap along the Thuadara-Sendhwa-Sindad profile across Narmada-Son lineament, India. In: R.F. Mereu, S. Muller and D.M. Fauntain (eds): *Properties of earth's lower crust*. American Geophysical Union Monograph 51, 127–141.

Kalina, A.I. 1983. Combined cycle and waste heat recovery power systems based on a novel thermodynamic energy cycle utilizing low-temperature heat for power generation. *Proceedings of the 1983 Joint Power Generation Conference*, Indianapolis, IN, 1983. ASME Paper No. 83-JPGC-GT-3, American Society of Mechanical Engineers, New York, NY, 1983, pp. 1–5.

Kalina, A.I. and Leibowitz, H.M. 1987. Applying Kalina technology to a bottoming cycle for utility combined cycles. *Proceedings of the Gas Turbine Conference and Exhibition*, 31 May–June 4, 1987, Anaheim, CA, ASME Paper No. 87-GT-35.

Kalina, A.I. 2006. New thermodynamic cycles and power systems for geothermal applications. Trans. GRC, 30, 2006. 747–750.

Kaila, K.L., Rao, I.B.P., Koteswar Rao, P., Madhava Rao, Krishna, V.G. and Sridhar, A.R. 1981. DSS studies over Deccan trap along the Thuadara-Sendhwa-Sindad profile across

Narmada-Son lineament, India. In: R.F. Mereu, S. Muller and D.M. Fauntain (eds): *Properties of earth's lower crust*. American Geophysical Union Monograph 51, 127–141.

Kasameyer, R. 1997. Brief Guidelines for the development of inputs to CCTS from the technology working group. Working draft, Lawrence Livermore Laboratory, Livermore, California.

Karameldin, A. and Mekhermar, S. 2001. Sitting assessment of a water electricity cogeneration nuclear power plant in Egypt. *Desalination*, 137, 45–51.

Katzir, Y., Eyal, M., Litvinovsky, B.A., Jahn, B.M., Zanvilevich, A.N., Valley, J.W., Beeri, Y., Pelly, I. and Shimshilashvili, E. 2007. Petrogenesis of A-type granites and origin of vertical zoning in the Katharina pluton, Gebel Mussa (Mt. Moses) area, Sinai, Egypt. *Lithos*, 95: 208–228.

Kana, J.D., Djongyang, N., Raidabdi, D., Nouck, P.N. and Dadje, A. 2015. A review of geophysical methods for geothermal exploration. *Renew. Sustain. Energy Rev.*, 44, 87–95.

Kipsang, C. 2013. Cost model for geothermal wells. UN University Geothermal Training programme report 11, 23p.

Koulakov, I., El Khrepy, S., Al Arifi, N., Kuznetsov, P. and Kasatkina, E. 2015. Structural cause of a missed eruption in the Harrat Lunayyir basaltic field (Saudi Arabia) in 2009. *Geology*, 43, 395–398.

Koulakov, I., El Khrepy, S., Al Arifi, N., Sychev, I. and Kuznetsov, P. 2014. Evidence of magma activation beneath the Harrat Lunayyir basaltic field (Saudi Arabia) from attenuation tomography. *Solid Earth*, 5, 873–882.

Kohlani, T. 2013. Geothermal prospecting by geochemical methods in the Quaternary volcanic province of Dhamar (central Yemen). *J. Volcanol. Geother. Res.* 249, 95–108.

Kulkarni, S.G., Vijayanand, P., Aksha, M., Reena, P. and Ramana, K.V.R. 2008. Effect of dehydration on the quality and storage stability of immature dates (Pheonix dactylifera). *LWT Food Sci. and Tech*, 41, 278–283.

Krane, J. 2015. a refined approach: Saudi Arabia moves beyond crude. *Energy Policy*, 82, 99–104.

Kroner, A. and Stern, R.J. 2004. Pan African Orogeny in Encyclopedia of Geology: Vol. 1, Africa.

Kuster, D. 2009. Granitoid-hosted Ta mineralization in the Arabian-Nubian Shield: Ore deposit types, tectono-metallogenetic setting and petrogenetic framework. *Ore geology reviews*, 35, 68–86.

Lahitte, P., Gillot, P.Y., Kidane, T., Courtillot, V. and Bekele, A. 2003. New age constraints on the timing of volcanism in central Afar, in the presence of propagating rifts. *J. Geophy. Res.*, 108, doi:10.1029/2001JB001689, 2003.

Langelier, W. and Ludwig, H. 1942. Graphical methods for indicating the mineral character of natural waters. *J. Am. Water Assoc.* 16, 141–164.

Lashin, A. 2012. A preliminary study on the potential of the geothermal resources around the Gulf of Suez, Egypt. *Arabian J. Geosci.* DOI 10.1007/s12517-012-0543-4.

Lashin, A. and Al Arifi, N. 2012. The geothermal potential of Jizan area, Southwestern parts of Saudi Arabia. *International J. Physical Sci.* 2012; 7:664–675.

Lashin, A. 2015. Geothermal Resources of Egypt: Country Update, Proceedings, World Geothermal Congress 2015, Melbourne, Australia, International Geothermal Association, 13 p.

Lazzeri, L. 1997. Application of Kalina cycle as bottoming cycle for existing geothermal plants. *Proceedings of the Florence World Energy Research Symposium*, July 30–August 1, 1997, Florence, Italy, 389–396.

Lashin, A., Chandrasekharam, D., Al Arifi, N., Al Bassam, A. and Chandrasekhar, V. 2014a. Geothermal energy resources of wadi Al-Lith, Saudi Arabia. *J. African Earth. Sci.* 97, 357–367.

Lashin, A., Al Arifi, N., Chandrasekharam, D., Al Bassam, A., Rehman, S. and Pipan, M. 2015. Geothermal Energy Resources of Saudi Arabia: Country Update. Proceed. World Geothermal Congress 2015, Melbourne, Australia, 2015 CD.

Le Fort, P. and Rai, S.M. 1999. Pre-Tertiary felsic magmatism of the Nepal Himalaya: recycling of continental crust. *J. Asian Earth Sciences* 17, 607–628.

Lachenbruch, A.H. 1968. Preliminary geothermal model of the Sierra Nevada. *Journal of Geophysical Research* 1968; 73:6977–6989.

Le Pichon, X. and Gaulier, J. M. 1988. The rotation of Arabia and the Levant fault system. *Tectonophy*. 153, 271–294.

Lowensterna, J.B., Janika, C.J., Fournier, R.O., Tesfaib, T., Duffield, W.A., Clynne, M.A., Smith, J.G., Woldegiorgis, L., Weldemariam, K. and Kahsai, G. 1999. A geochemical reconnaissance of the Alid volcanic center and geothermal system\Danakil depression, Eritrea. *Geothermics* 28, 161–187.

Lund, J.W. 2002. Direct utilization of geothermal resources, in "Geothermal Energy Resources for Developing countries" (Eds) D. Chandrasekharam and J. Bundschuh, A.A. Balkema Press, The Netherlands, 412 p.

Lund, J.W. and Lienau, P.J. 2002. Agri business uses of geothermal energy, in Geothermal energy resources for developing countries (Eds) D. Chandrasekharam and J. Bundschuh, A A Balkema Press, The Netherlands, 412 p.

Lund, J.W., Freeston, D.H. and Boyd, T.L. 2010. Direct Utilization of Geothermal Energy 2010 Worldwide Review, Proceedings World Geothermal Congress 2010, Bali, Indonesia, 25–29 April 2010, 1–23.

Lund, J. and Boyd, T. 1999. Small geothermal power projects examples. *Geo Heat Centre Bulletin*, 20, 9–26.

Lund, J. W. and Boyd, T. 2015. Direct utilization of geothermal energy 2015 World review. *Proceedings World Geothermal Congress*, 2015.

Lund, J.W. 1996. Lectures on direct utilization of geothermal energy. Geothermal training programme, United Nations University, Iceland, 124 p.

Lundmark, A.M., Andresen, A., Hassan, M.A., Augland, L.E. and Boghdady, G.Y. 2012. Repeated magmatic pulses in the East African Orogen in the Eastern Desert, Egypt: An old idea supported by new evidence. *Gondwana Research* 22, 227–237.

Lyberis, N. 1988. Tectonic evolution of the Gulf of Suez and the Gulf of Aqaba. *Tectonophysics*, 153, 209–220.

Makris, J. and Rihm, R. 1991. Shear controlled evolution of the Red Sea: pull apart model. *Tectonophy* 198, 441–466.

Mattash, M.A. 1994. Study of the Cenozoic volcanics and Their Associated Intrusive Rocks in Yemen in Relation to Rift Development. Ph.D. Thesis, Univ. Budapest, Hungarian Academy of Sciences.

Marchesini, E., Pistolesi, A. and Bolognini, M. 1962. Fracture pattern of the natural steam area of Larderello, Italy from air photographs. Sym. Photo interpretation, Delft. 524.

McCombe, D.A., Fernette, G.L. and Aalawi, A.J. 1994. The geological and mineral resources of Yemen. Ministry of oil and Mineral Resourcs, Geological Survey of Yemen ed., Sana'a (Yemen), 128 pp.

McKenzie, D.P., Davies, D. and Molnar, P. 1970. Plate tectonics of the Red Sea and East Africa. Nature, 26, McKenzie, D.P., Davies, D. and Molnar, P. Molnar P. Plate tectonics of the Red Sea and East Africa. Nature, 26, 243–249.

Makris, J., Zimmermann, J., Balan, A. and Lebras, A. 1975. Gravity study of the Djibouti area. *Tectonophy*. 27, 177–185.

Mezher, T., Fath, H., Abbas, Z. and Khaled, A. 2011. Techno-economic assessment and environmental impacts of desalination technologies. *Desalination*. 266, 263–273.

Menzies, M., Al-Kadasi, M.A., Al-Khirbash, S., Al-Subbary, A., Baker, J., Blakey, S., Bosence, D., Davison, I., Dart, C., Owen, L., McClay, K., Nichols, G., Yelland, A., Watchorn, F., 2001. Geology of Yemen. In: Geology and Mineral Resources of Yemen, Published by the Geological Survey and Mineral Resources of Yemen, Sana'a, pp. 21–48.

Menzies, M., Bosence, D., El Nakhal, H.A., Al Khirbash, S., Al Khadasi, M.A., Al Subbary, A., 1990. Lithospheric extension and the opening of the Red Sea: sediment-basalt relationships in Yemen. *Terra Nova* 2, 340–350.

Minissale, A. 1991. The Larderello geothermal field: A review. *Earth Science Reviews* 31, 133–151.

Minissale, A., Vaselli, O., Chandrasekharam, D., Magro, G., Tassi, F. and Casiglia, A. 2000. Origin and evolution of "intracratonic" thermal fluids from central western peninsular India. *Earth and Planetary Science Letters* 181, 377–394.

Minissale, A., Chandrasekharam, D., Vaselli, O., Magro, G., Tassi, F., Pansini, G.L. and Bhramhabut, A. 2003. Geochemistry, geothermics and relationship to active tectonics of Gujarat and Rajasthan thermal discharge, India. *J. Volcanology and Geothermal Research* 127, 19–32.

Minissale, A., Mattash, M.A., Vaselli, O., Tassi, F., Al-Ganad, I.N., Selmo, E., Shawki, N.M., Tedesco, D., Poreda, R., Ad-Dukhain, A.M. and Hazzae, M.K. 2007. Thermal springs, fumaroles and gas vents of continental Yemen: Their relation with active tectonics, regional hydrology and the country's geothermal potential. *Applied Geochem.*, 22, 799–820.

Minissale, A., Vaselli, O., Mattash, M., Montegrossi, G., Tassi, F., Ad-Dukkain, A., Kalberkamp, U., Al-Sabri, A. and Al Kohlani, T. 2013. Geothermal prospecting by geochemical methods in the Quaternary volcanic province of Dhamar (central Yemen). *J. Volcanol. Geother. Res.* 249, 95–108.

Minster, J.B. and Jordan, T.H.: Present-day plate motions. *J. Geophysical Research* 83 (1978), pp. 5331–5334.

Milkereit, B. and Fluh, E.R. 1985. Saudi Arabian refraction profile: crustal structure of the Red Sea Arabian shield transition. *Tectonophy.* 111, 283–298.

MEE, 2009. National strategy for renewable energy and energy efficiency, Ministry of Electricity and Energy (MEE), Yemen Republic, Strategy report, 11 p.

MIT, 2006. The Future of Geothermal Energy – Impact of Enhanced Geothermal Systems (EGS) on the United States in the 21st Century, An assessment by an MIT led interdisciplinary panel, MIT – Massachusetts Institute of Technology, Cambridge, MA. (2006), 358 p.

Mitchell, A.H.G. 1993. Cretaceous-Cenozoic tectonic events in the western Myanmar (Burma)-Assam region. *J. Geological Society, London* 150, 11,089–11,102.

Moufti, M.R., Moghazi, A.M. and Ali, K.A. 2013. $^{40}Ar/^{39}Ar$ geochronology of the Neogene-Quaternary Harrat Ai Madinah intercontinental volcanic field, Saudi Arabia: Implications for duration and migration of volcanic activity. *J. Asian Earth Sci.*, 62, 253–268.

Mooney, W.D., Gettings, M.E., Blank, H.R. and Healy, J.H. 1985. Saudi Arabian seismic refraction profile: A travel time interpretation of crustal and upper mantle structure. *Tectonophy.* 111, 173–246.

Morgan, P., Blackwell, D.D., Fanis, T.G., Boulos, F.K. and Salib, P.G. 1976. Preliminary temperature gradient and heat flow values for northern Egypt and the Gulf of Suez from oil well data. In: Proceedings International Congress on Thermal Waters. Geothermal Energy and Volcanism of Mediterranean Area, vol. I, *Geothermal Energy*, pp. 424–438.

Morgan, P. and Swanberg, C.A. 1978. Heat Flow and the Geothermal Potential of Egypt. *Pageophl.* 17:213–226.

Morgan, P., Blackwell, D. D., Farris, J. C., Boulos, F. K. and Salib, P. G. 1977. Preliminary geothermal gradient and heat flow values for northern Egypt and the Gulf of Suez from oil well data, in Proceedings, Int. Cong. Thermal Waters, Geothermal Energy and Vulcanism of the Mediterranean Area, Nat. Tech. Univ. Athens, Greece. v. I, 424–438.

Morgan, P., Boulos, F.K., Hennin, S.F., El-Sherif, A.A., EI-Saycd, A.A., Basta, N.Z. and Melek, Y.S. 1981. Geophysical investigations of a geothermal anomaly at Wadi Ghadir, eastern Egypt. in Proceedings. First Annual Meeting Egyptian Geophysical Soc., Cairo, Sept. 28, 29: 17 pp. + 6 figures.

Morgan, P., Boulos, F. K. and Swanberg, C. A. 1983. Regional geothermal exploration in Egypt, Geophysical Prospecting, 31: 361–376.

Morgan, P., Boulos, F.K., Hennin, S.F., El-Sherif, A.A., El-Syed, A.A., Basta, N.Z. and Melek, Y.S. 1985. Heat flow in eastern Egypt: The thermal signature of a continental breakup. *J. Geodynamics*, 4, 107–131.

Moore, J.M. 1979. Primary and secondary faulting in the Najd fault systeem, Kingdom of Saudi Arabia. USGS open file report 79-1661, 26p.

Moussa, E.M.M., Stern, R., Manton, W.I. and Ali, K.A. 2008. SHRIMP zircon dating and Sm/Nd isotopic investigations of Neoproterozoic granitoids, Eastern Desert, Egypt. Precambrian Research 160, 341–356. doi:10.1016/j.precamres.2007.08.006.

Moussa, O.A. and Suleiman, H. 2015. Country report, Geothermal development in Djibouti. Proceedings, World Geothermal Congress, April 19, 29, 2015, Melbourne, Australia.

Nagy, R.M., Ghuma, M.A., Rogers, J.J., 1976. A crustal suture and lineaments in North Africa. *Tectonophysics* 31, 67–72.

Nakicenovic, N. and Swart, R. 2000. IPCC Special Report on Emissions Scenarios: A special report of working group III of the Intergovernmental Panel on Climate Change, Cambridge University Press, Cambridge, URL: http://www.grida.no/climate/ipcc/emission/.

Naqvi, S.M., Divakar, Rao, V. and Han Narain. 1974. The protocontinental growth of the Indian shield and the antiquity of its rift valleys. *Precambrian Research*, 1, 345–398.

Naqvi, S.M., Divakar, Rao, V. and Han Narain. 1978. The primitive crust: Evidence from the Indian shield. *Precambrian Research* 6 (1978): pp. 323–345.

Nassif, A.A. 2012. Renewable and efficiency energy initiatives in Yanbu industrial city. In: Fourth Saudi Solar Energy Forum Riyadh, 8–9 May.

Nehlig, P., Genne, A., Asfirane, F., Lasserre, J.L., Le Goff, E., Nicol, N., Salpeteur, N., Shanti, M., Thieblemont, D. and Truffert, C. 2002. A review of the Pan-African evolution of the Arabian Shield. *GeoArabia*, 7, 103–123.

Nielson, D. 1989. "Competitive economics of Geothermal Energy: The Exploration and Development Perspective", in "Geothermal Development Opportunities in Developing Countries" US Department of Energy Publication 1989. 41–48.

Novakov, T., Ramanathan, V., Hansen, J.E., Kirchstetter, T.W., Sato, M., Sinton, J.E. and Sathaye, J.A. 2003. Large historical changes of fossil fuel black carbon aerosols. *Geophy. Research Lett.*, 30, 1324.

Nuti, S. 1991. Isotope techniques in geothermal studies *in* "Application of geochemistry in geothermal reservoir development" (Ed) Franco D' Amore, Series of technical Guides on the use of geothermal energy, UNITAR/UNDP Centre on Small Energy Resources. 411p.

NWA, 2015. Nuclear World Association: http://www.world-nuclear.org/info/Country-Profiles/Countries-O-S/Saudi-Arabia/ (accessed on 29 May 2015).

OPEC 2014. Annual Statistical Bulletin. 112 p.

Ouda, O.K.M., Raza, S.A., Al-Waked, R., Al-Asad, J.F. and Nizami, A.S. 2015. Waste to energy potential in the western province of Saudi Arabia. *J. KSU-Engg. Sci.* (in press).

Omenda, P., Varun, C. and Chandrasekhram, D. 2012. High heat generating granites of Uganda and Tanzania: Possible EGS sources in Eastern Africa. *Proceed. 4th African Rift Geothermal Conference, Nairobi, Kenya (CD).*

Patchett, P.J. and Chase, C.G. 2002. Role of transform continental margin in major crustal growth episode. Geology, 30:39–42.

Page, B.G.N., Bennet, J.D., Cameron, N.R., Bridge, D., Mc Jeffery, D.H., Keats,W. and Thaib, J. 1979. A review of the main structural and magmatic features of northern Sumatra. *J. Geological Society London.* 136, 569–579.

Pallister, J.S. 1986. Red Sea rift magmatism near Al Lith, Kingdom of Saudi Arabia. USGS Open file report 86-565.

Pallister, J.S. 1987. Magmatic history of Red Sea rifting: Prospective from the central Saudi Arabian coastal plain. *Bul. Geol. Soc. Am.*, 98, 400–417.

Pallister, J.S., McCausland, W.A., Jonsson, S., Lu, Z., Zahran, H.M., El Hadidy, S., Aburukbah, A., Stewart, I. C. F., Lundgren, P.R., White, R.A. and Moufti, M.R.H. 2010. Broad accommodation of rift-related extension recorded by dyke intrusion in Saudi Arabia. *Nature Geoscience* 3, 705–712. doi:10.1038/ngeo966.

Patton, T.L., Moustafa, A.R., Nelson, R.A., Abdine, A.S., 1994. Tectonic evolution and structural setting of the Suez rift. In: Landon, S.M. (Ed.), Interior Rift Basins, vol. 58. *Am. Assoc. Pet. Geol. Mere*, pp. 9–55.

Parker, R. The Rosemanowes HDR project 1983–1991. 1999. Geothermics, 28, 603–615.

Park, Y., Nyblade, A., Rodgers, A. and Al-Amri, A. 2008. S wave velocity structure of the Arabian Shield upper mantle from Rayleigh wave tomography. *Geochemistry Geophysice Geosystems* 9, 1–15.

Perrin, M., Saleh, A. and Valdivia, L.A. 2009. Cenozoic and Mesozoic basalt from Egypt: a preliminary survey. *Earth Planets Space*, 61, 51–60.

Potter, R.M., Smith, M.C. and Robinson, E.S. 1974. "Method of extracting heat from dry geothermal reservoirs," U. S. patent No. 3,786,858.

Plakfer, G., Agar, R., Asker, A.H. and Haulf, M. 1987. Surface effects and tectonic setting of the 13th December 1982 north Yemen earthquake. Bulletin of the Seismological Society of America 77, 2,018–2,037.

Pruess, K. 2006. Enhanced geothermal systems (EGS) using CO_2 as working fluid – A novel approach for generating renewable energy with simultaneous sequestration of carbon. *Geothermics*, 35, 351–367.

Rabie, S.I., Assaf, H.S., El Sayed, M., El Kattan and Osman, H.S. Kattan, 1996. A exploration in El Missikat prospect area, central eastern desert, Egypt. *Radiat Phy Chem.* 47, 769–774.

Raslan, M.F., El-Feky, M.G., 2012: Radioactivity and mineralogy of the altered granites of the Wadi Ghadir shear zone, South Eastern Desert, Egypt. *Chin. J. Geochem.*, 31:030–040.

Razzano, F. and Cei, M. 2015. Geothermal Power Generation in Italy 2010–2014 Update Report. Proceed. World Geothermal Congress, 2015, Melbourne, Australia, April 19–24.

Riki, F.I., Fauzi, A. and Suryadarma. 2005. The progress of geothermal energy resources activities in Indonesia. Country update, World Geothermal Congress, Turkey 2005.

Rybach, L. 1976. Radioactive Heat Production: A Physical Property Determined by the Chemistry *in* R.G.I. Strens (Etd) "The Physical and Chemistry of Minerals and Rocks". Wiley-Interscience Publication, New York, 245–276.

Ramanathan, V. and Carmichael, G. 2008. Global and regional climate change due to black carbon. *Nature Geoscience*, Nature Publishing Group, 1, 220–228.

Ramli, M.A.M., Hiendro, A., Sedraoui, K. and Twaha, S. 2015. Optimal sizing of grid connected photovoltaic energy system in Saudi Arabia. *Renew. Energy*. 75, 489–495.

Robinson, S., Strzepek, K. and Cervigni, R. 2013. The Cost of Adapting to Climate Change in Ethiopia: Sector-Wise and Macro-Economic Estimates: Ethiopia Strategy Support Programme II. Inter. Food Policy Res. Institute ESSP working paper 53, 26p.

Rogers, J.J.W., Miller, J.S., Climents, S.A., 1995. A Pan-African zone linking East and West Gondwana. *Memoir of the Geological Society of India* 34, 11–23.

Saleh, G.M. 2006. Uranium mineralization in the muscovite-rich granites of the Shalatin region, Southeastern Desert, Egypt. *Chinese Journal of Geochemistry*, 25(1): 1–15.

Saleh, S., Jahr, T., Jentzsch, G., Saleh, A. and Ashour N.M.A. 2006. Crustal evaluation of the northern Red Sea rift and Gulf of Suez, Egypt from geophysical data: 3-dimensional modelling. *J. African Earth Sci.* 45, 257–278.

Saleh, G.M., Mostafa, D.A., Darwish, M.E. and Salem, I.A. 2014. Gabal El Faliq Granitoid rocks of the Southeastern Desert, Egypt: Geochemical constraints, mineralization and

Spectrometric Prospecting. *Standard Global Journal of Geology and Explorational Research*, 1(1): 009–026.

Sanyal, S. 2004. Cost of geothermal power and factors that affect it. *Proceed. 29th workshop on geothermal reservoir engg. Stanford, California, 26–28* (SGP TR 175).

Schandelmeier, H. and Pudlo, D. 1990. The Central-African fault zone in Sudan–a possible continental transform fault. *Berl. Geowiss. Abh.* 120-A, 31–44.

Sharqawy, M.H., Said, S.A., Mokheimer, E.M., Habib, M.A., Badr, H.M. and Al-Shayea, N.A. 2009. First insitu determination of the ground thermal conductivity for borehole heat exchanger applications in Saudi Arabia. *Renewable Energy*, 34, 2218–2223.

Scheiber, J., Seibt, A., Birner, J., Genter, A., Cuenot, N. and Moeckes, W. 2015. Scale Inhibition at the Soultz-sous-Forêts (France) EGS Site: Laboratory and On-Site Studies. *Proceed. World Geothermal Congress 2015, Melbourne, Australia.*

Schneider, D.A., Edwards, M.A., Zeitler, P.K. and Coath, C.D. 1999a. Mazeno Pass pluton and Jutial pluton, Pakistan Himalaya: age and implications for entrapment mechanism of two granites in Himalayas. *Contributions to Mineralogy Petrology* 136, 273–284.

Schneider, D.A., Edwards, M.A., Kidd, W.S.F., Asif Khan, M., Seeber, L. and Zeitler, P.K. 1999b: Tectonics of Nanga Parbat, western Himalaya: Synkinamatic plutonism within the doubly vergent shear zones of a crustal-scale pop-up structure. *Geology* 27, 999–1002.

Schneider, D.A., Edwards, M.A., Kidd, W.S.F., Zeitler, P.K. and Coath, C.D. 1999c. Early Miocene anatexis identified in the western syntaxis, Pakistan Himalaya. *Earth and Planetary Science Letters* 167, 121–129.

Schneider, D.A., Edwards, M.A., Zeitler, P.K. and Coath, C.D. 1999a. Mazeno Pass pluton and Jutial pluton, Pakistan Himalaya: age and implications for entrapment mechanism of two granites in Himalayas. *Contributions to Mineralogy Petrology* 136, 273–284.

Schneider, D.A., Edwards, M.A., Kidd, W.S.F., Asif Khan, M., Seeber, L. and Zeitler, P.K. 1999b. Tectonics of Nanga Parbat, western Himalaya: Synkinamatic plutonism within the doubly vergent shear zones of a crustal-scale pop-up structure. *Geology* 27, 999–1002.

Schneider, D.A., Edwards, M.A., Kidd, W.S.F., Zeitler, P.K. and Coath, C.D. 1999c. Early Miocene anatexis identified in the western syntaxis, Pakistan Himalaya. *Earth and Planetary Science Letters* 167, 121–129.

Searle, M.P. 1999a. Extensional and compressional faults in the Everest massif, Khumbu Himalayas. *J. Geological Society London* 156, 227–240.

Searle, M.P. 1999b. Emplacement of Himalayan leucogranites by magma injection along giant complexes: examples from the Cho Oyu, Gyachung Kang and Everest leuco-granites (Nepal Himalaya). *J. Asian Earth Sciences* 17, 773–783.

Sebai, A., Zumbo, V., Féraud, G., Bertrand, H., Hussain, A.G., Giannerini, G. and Campredon, R., 1991. 40Ar/39Ar dating of alkaline and tholeiitic magmatism of Saudi Arabia related to the early Red Sea Rifting. Earth and Planetary Science Letters 104, 473–487.

Segar, C. 2014. Renewable augment gas Saudi energy mix. J. IEA 7, 40e41.

Singh, H.K., Chandrasekharam, D., Vaselli, O., Trupti, G., Singh, B., Lashin, A. and Al Arifi, N. 2015. Physico chemical characteristics of Jharkhand and west Bengal thermal springs along SONATA mega lineament, India. *J. Earth Syst. Sci.* **124**, 419–430.

Simons, G., Peterson T. and Poore, R. 2001. "*California Renewable Technology Market and Benefits Assessment*", Electric Power Research Institute, 2001.

Smith, M.C. 1995. The furnace in the basement Part I: The early days of the ot Dr Rock geothermal energy programme: 1970–1973. Los Alamos National Laboraory, report 12809, Part I, 229p.

Stauffer, P.H. 1983. Unraveling the mosaic of Palaeozoic crustal blocks in southeast Asia. *Geologische Rundschau* 72, 1061–1080.

Stefansson, V. 1992. Success in geothermal development. *Geothermics* 21, 823–834.

Stefansson, V. 2002. Investment cost for geothermal power plants. *Geothermics*, 31, 263–272.

Stern, R.J., Gottfried, D. and Hedge, C.E. 1984. Late Precambrian rifting and crustal evolution in the northeast Desert of Egypt. *Geology* 12, 168–172.

Stern, R.J. 1985. The Najd Fault System, Saudi Arabia and Egypt: a late Precambrian rift-related transform system? *Tectonics* 4, 497–511.

Stern, R.J. 1994. Arc assembly and continental collision in the Neoproterozoic East African Orogen: implications for the consolidation of Gondwanaland. *Annu. Rev. Earth Planet. Sci.* 22, 319–351.

Stern, R.J. and Johnson, P. 2010. Continental lithosphere of the Arabian Plate. A geologic, petrologic, and geophysical synthesis. Earth Sci. Rev. 101, 29–67.

Stoeser, D.B. and Camp, V.E. 1985. Pan-African microplate accretion of the Arabian shield: Geological Society of America Bulletin, 96, 817–826.

Shaahid, S.M. and Elhadidy, M.A. 1994. Wind and solar energy at Dhahran, Saudi Arabia. Renew. Energy, 4, 441–445.

Shaahid, S.M., Al Hadhrami, L.M. and Rahman, M.K. 2013. Economic feasibility of development of wind power plants in coastal locations of Saudi Arabia- a review. RSER, 19, 589–597.

Schmitt, R.C. 1981. Agriculture, Greenhouses, Wetland and Other Beneficial Uses of Geothermal Fluids and Heat. *Proceedings of the First Sino/US. Geothermal Resources Conference (Tianjin, PRC)*, Geo-Heat Center, Klamath Falls, OR.

Streets, D.G., Bond, T.C., Lee, T. and Jang, C. 2004. On the future of carbonaceous aerosol emissions. *J. Geophy. Res.*, 109, D 24212, doi: 10.1029/2004JD004902, 2004.

Stern, R.J. 1994. Arc assembly and continental collision in the Neoproterozoic East African Orogen-Implications for the consolidation of Gondwanaland. *Earth. Planet. Sci. Ann. Rev.* 22, 319–351.

Stoeser, D.B. 1986. Distribution and tectonic setting of plutonic rocks of the Arabian Shield. *J. African Earth Sci.*, 4, 31–46.

Swanberg, C.A., Morgan, P. and Boulos, F.K. 1983. Geothermal potential of Egypt. *Tectonophy.* 77, 94.

Tenma, N., Iwakiri, S.I. and Matsunaga, I. 1998. Development of hot dry rock technology at Hijiori test site: Programme for a long term circulation test. *Energy Sources*, 20:8, 753–762, DOI: 10.1080/00908319808970095.

Tenma, N., Yamaguchi, T., Oikawa, Y. and Zyvoloski, G. 2001. "Comparison of the Deep and the Shallow Reservoirs at the Hijiori HDR Test Site using FEHM Code." *Proc. 26th Stanford Geothermal Workshop*, CA.

Tenma, N., Yamaguchi, T. and Zyvoloski, G. 2008. The Hijiori hot dry test site, Japan. Evaluation and optimization of heat extraction from a two layered reservoir. Geothermics, 37, 19–52.

Thorhallsson, S. and Sveinbjornsson, B.M. 2012. Geothermal drilling cost and drilling effectiveness. Lecture at the short course on "Geothermal Development and Geothermal wells" organized by UNU-GTP and LaGeo, Santa Tecla, El salvador, March 11–17, 2012.

Tonani, F. 1980. Some remarks on the application of geochemical techniques in geothermal exploration. *Proceedings 2nd Symposium on Advance European Geothermal Resources*, Strasbourg, 428–443.

Truesdell, A.H. and Hulston, J.R. 1980. Isotopic evidence on environments of geothermal systems *in* Handbook of Environmental Isotope Geochemistry (Eds) P. Fritz and J. Ch. Fontes, Elsevier Pub. Co., Amsterdam, 183–230.

Truesdell, A.H. 1976. Summary of section III geochemical techniques in exploration. *Proceedings Second United Nations Symposium on the development and use of geothermal resources*, Vol. 1., San Francisco, Washington, DC.

Tucho, G.T., Weesie, D.M. and Nonhebel, S. 2014. Assessment of renewable energy resources potential for large scale and standalone applications in Ethiopia. *Renew. Sustainable Energy Rev.* 40, 422–431.

Taleb, H.M. and Sharples, S. Developing sustainable residential building in Saudi Arabia: A case study. *App. Energy* 2011; 88:383–391.
Tassew, M. 2015. Expansion work and experience gained in operation of Aluto Langano geptheral power plant. Proceedings World Geothermal Comgress 2015, Australi, April 19–25. Proceedings World Geothermal Congress 2015.
U N D P, 2014. African Economic Outlook 2014, 317p.
UNFCC, 2006. UNFCCC: United Nations framework convention for climate change, synthesis report.
Ueipass, M.W. and Sylvie, L. 2012. Architecture of rifted continental margins and breakup evolution: Insights from the south Atlantic, north Atlantic and Reed Sea-Gulf of Aden conjugate margins. *Geol. Sco. London*. 369, 1–40.
USDA, 2013. Global Agricultural Information Network and Feed annual: Saudi Arabia, USDA foreign agricultural service report SA-1302:18.
USGS, 2002. US Geological Survey Minerals year book 2002: The Mineral industries of Djibouti, Eretria and Somalia. 9p.
U.S. Central Intelligence Agency, 2000, Djibouti, *in* World factbook 2000: U.S. Central Intelligence Agency, pp. 139–140.
USDA, 2013. Global Agricultural Information Network and Feed annual: Saudi Arabia, USDA foreign agricultural service report SA, 1302:18.
Valdimarsson, P. and Eliasson, L. 2003. Factor influencing the economics of the Kalina power cycle and situations of superior performance. *Proceedings International Geothermal Conference*, 14–17 September 2003, Reykjavik, Iceland, 33–40.
Varun, C. and Chandrasekharam, D. 2009. Geothermal Systems in India. *Geothermal. Res. Council Trans*, 33, 607–610.
Vernier, R., Laplaige, P., Desplan, A. and Boissavy, C. 2015. France Country Update. Proceed. World Geothermal Congress, 2015, Melbourne, Australia.
Vernekar, A.D. 1975. A calculation of Normal temperature at the Earth's surface. *J. Atmospheric Sciences*. 32:2067–2081.
Vimmerstedt, L. 1998. Opportunities for small geothermal projects: Rural power for Latin America, the Caribbean and the Philippines. National Renewable Energy Laboratory report. NREL/TP-210-25107, 66p.
Vijayaraghavan, S. and Goswami, D.Y. 2005. Organic working fluids for a combined power and cooling cycle. *J. Energy Resources Technology*, 127, 125–30.
Vimmerstedt, L. 2002. Small geothermal projects for rural electrification. In "Geothermal resources for developing countries" (Eds) D. Chandrasekharam and J. Bundschuh, A.A. Bakema Pub., 103–129.
Vijayaraghavan, S. and Goswami, D.Y. 2005. Organic working fluids for a combined power and cooling cycle. *J. Energy Resources Technology* 127, 125–30.
Weber, J., Ganz, B., Schellschmidt, R., Sanner, B., and Schulz, R. 2015. Geothermal energy use in Germany. World Geothermal Congress, Melbourne, Australia, 2015.
Waibel, A.F., Frone, Z. and Jaffe, T. 2012. Geothermal Exploration at Newberry Volcano, Central Oregon. GRC *Transactions*, Vol. 36, 803–810.
WFP, 2010. World Food Programme, Market study, Yemen. 64p.
Woldegabriel, G., Aronson, J. and Walter, R. 1990. Geology, geochronology, and rift basin development in the central sector of the Main Ethiopian rift, *Geol. Soc. Am. Bull.* 102, 439–458.
WB, 2012. Project performance assessment report – Republic of Djibouti: Food crisis response development policy grant. 56p.
WB, 1996. Rural Energy and Development: Improving Energy Supplies for Two Billion People. The World Bank Report, Washington, D.C. 43p.
WB. 2014. World Bank Development Indicators, WB: http://data.worldbank.org/data-catalog/world-development-indicators (accessed on 1 Sept 2014).

Wilson, J.T. 1968. Static or mobile earth: the current revolution. *Proc. Am. Philos. Soc.*, 112, 309–320.

Worell, E., Prince, L., Martin, N., Hendriks, C. and Meida, L.O. 2001. Carbon dioxide emissions from the global cement industry. *Annul Rev. Energy and Environ.*, 26, 303–329.

Wolfenden, E., Ebinger,C., Yirgu, G., Deino, A. and Ayalew, D. 2004. Evolution of the northern Main Ethiopian rift: birth of a triple junction. *Earth. Planet. Sci. Lett.* 224, 214–228.

Yohannes, E. 2007. Geothermal exploration in Eritrea. Status report and discussion. Proceed. Surface Exploration for Geothermal Resources, organized by UNU-GTP and KenGen, Lake Naivasha, Kenya, 2–17 November.

Zan, L., Gianelli, G., Passerini, P., Troisi, C. and Haga, A.O. 1990. Geothermal exploration in the Republic of Djibouti: Thermal and geological data of the Hanle and Assa areas. *Geothermics*, 19, 6, 561–582.

Zaher, M.A., Saibi, H., El-Nouby, M., Ghamry, E. and Ehara, S. 2011. A preliminary regional geothermal assessment of the Gulf of Suez, Egypt. *J. African Earth Sci.* 60, 117–132.

Zaher, A.M., Saibi, H. and Ehara, S. 2012. Geochemical and stable isotopic studies of Gulf of Suez's hot springs, Egypt *Chin. J. Geochem.* 31, 120–127.

Zlotnicki, M.M.J. 2001. Fluid circulation in the active emerged Asal rift (east Africa, Djibouti) inferred from self potential and telluric-telluric prospecting. *Tectonophy.* 339, 455–472.

Zaher, M.A., Saibi, H., Nishijim, J., Fujimitsu, Y., Mesbah, H. and Ehara, S. 2012. Exploration and assessment of the geothermal resources in the Hammam Faraun hot springs, Sinai Peninsula, Egypt. *J. Asian Erth Sci.*, 45, 256–267.

Subject Index